Chemistry
Essentials
FOR
DUMMIES®

by John T. Moore, EdD

WILEY

John Wiley & Sons, Inc.

Chemistry Essentials For Dummies®
Published by
John Wiley & Sons, Inc.
111 River St.
Hoboken, NJ 07030-5774
www.wiley.com

Copyright © 2010 by John Wiley & Sons, Inc., Hoboken, New Jersey

Published by John Wiley & Sons, Inc., Hoboken, New Jersey

Published simultaneously in Canada

For general information on our other products and services, please contact our Customer Care Department within the U.S. at 877-762-2974, outside the U.S. at 317-572-3993, or fax 317-572-4002.

For technical support, please visit www.wiley.com/techsupport.

Wiley publishes in a variety of print and electronic formats and by print-on-demand. Some material included with standard print versions of this book may not be included in e-books or in print-on-demand. If this book refers to media such as a CD or DVD that is not included in the version you purchased, you may download this material at http://booksupport.wiley.com. For more information about Wiley products, visit www.wiley.com.

Library of Congress Control Number: 2010925163

ISBN 978-0-470-61836-3 (pbk); ISBN 978-0-470-64449-2 (ebk); ISBN 978-0-470-64450-8 (ebk); ISBN 978-0-470-64451-5 (ebk)

Manufactured in the United States of America

15

About the Author

John T. Moore grew up in the foothills of Western North Carolina. He attended the University of North Carolina-Asheville where he received his bachelor's degree in chemistry. He earned his Master's degree in chemistry from Furman University in Greenville, South Carolina. After a stint in the United States Army, he decided to try his hand at teaching. In 1971, he joined the chemistry faculty of Stephen F. Austin State University in Nacogdoches, Texas where he still teaches chemistry. In 1985, he started back to school part time and in 1991 received his Doctorate in Education from Texas A&M University.

John's area of specialty is chemical education, especially at the pre-high school level. For the last several years, he has been the co-editor (along with one of his former students) of the *Chemistry for Kids* feature of *The Journal of Chemical Education*. He has authored *Chemistry For Dummies* and *Chemistry Made Simple*, and he's co authored *5 Steps To A Five: AP Chemistry*, *Chemistry for the Utterly Confused*, and *Biochemistry For Dummies*.

John lives in Nacogdoches, Texas with his wife Robin and their two dogs. He enjoys brewing his own beer and mead and creating custom knife handles from exotic woods. And he loves to cook. His two boys, Jason and Matt, remain in the mountains of North Carolina along with his twin grandbabies, Sadie and Zane.

Publisher's Acknowledgments

We're proud of this book; please send us your comments through our Dummies online registration form located at http://dummies.custhelp.com. For other comments, please contact our Customer Care Department within the U.S. at 877-762-2974, outside the U.S. at 317-572-3993, or fax 317-572-4002.

Some of the people who helped bring this book to market include the following:

Acquisitions, Editorial, and Media Development

Senior Project Editor: Tim Gallan

Acquisitions Editor: Lindsay Lefevere

Senior Copy Editor: Danielle Voirol

Technical Reviewer: Medhane Cumbay, Patti Smykal

Editorial Program Coordinator: Joe Niesen

Editorial Manager: Michelle Hacker

Editorial Assistants: Jennette ElNaggar, David Lutton, Rachelle Amick

Cover Photo: © iStock/bratan007

Cartoons: Rich Tennant (www.the5thwave.com)

Composition Services

Project Coordinator: Patrick Redmond

Layout and Graphics: Carrie A. Cesavice, Joyce Haughey

Proofreaders: Rebecca Denoncour, Sossity R. Smith

Indexer: Potomac Indexing, LLC

Publishing and Editorial for Consumer Dummies

    Kathleen Nebenhaus, Vice President and Executive Publisher

    Kristin Ferguson-Wagstaffe, Product Development Director

    Ensley Eikenburg, Associate Publisher, Travel

    Kelly Regan, Editorial Director, Travel

Publishing for Technology Dummies

    Andy Cummings, Vice President and Publisher, Dummies Technology/General User

Composition Services

    Debbie Stailey, Director of Composition Services

Contents at a Glance

Contents

Introduction

Congratulations on making a step toward discovering more about what I consider a fascinating subject: chemistry. For more than 40 years, I've been a student of chemistry. This includes the time I've been teaching chemistry, but I still consider myself a student because I'm constantly finding out new facts and concepts about this important and far-reaching subject.

Hardly any human endeavor doesn't involve chemistry in some fashion. People use chemical products in their homes — cleaners, medicines, cosmetics, and so on. And they use chemistry in school, from the little girl mixing vinegar and baking soda in her volcano to the Ivy League grad student working on chemical research.

Chemistry has brought people new products and processes. Many times this has been for the good of humankind, but sometimes it's been for the detriment. Even in those cases, people used chemistry to correct the situations. Chemistry is, as has been said many times, the central science.

About This Book

My goal with this book is to give you the really essential information and concepts that you would face in a first semester chemistry class in high school or college. I've omitted a lot of topics found in a typical chemistry textbook. This book is designed to give you the bare essentials.

Remember, this is a light treatment. If you want more, many other books are available. My favorite, naturally, is *Chemistry For Dummies*. I understand the author is really a great guy.

Conventions Used in This Book

Here are a couple of conventions you find in *For Dummies* books:

- ✔ I use *italics* to emphasize new words and technical terms, which I follow with easy-to-understand definitions.

- ✔ **Bold** text marks keywords in bulleted lists and highlights the general steps to follow in a numbered list.

In addition, I've tried to organize this book in approximately the same order of topics found in a one-semester general chemistry course. I've included some figures for you to look at; refer to them as you read along. Also, pay particular attention to the reactions that I use. I've attempted to use reactions that you may be familiar with or ones that are extremely important industrially.

Foolish Assumptions

I don't know your exact reasons for picking up this guide, but I assume you want to know something about chemistry. Here are some reasons for reading:

- ✔ You may be taking (or retaking) a chemistry class. This book offers a nice, quick review for your final exam. It can also give you a refresher before you plunge into a new course, such as biochemistry or organic chemistry.

- ✔ You may be preparing for some type of professional exam in which a little chemistry appears. This book gives you the essentials, not the fluff.

- ✔ You may be a parent trying to help a student with his or her homework or assignment. Pay attention to what your child is currently studying and try to stay a little ahead.

- ✔ Finally, you may be what people call a "nontraditional student." You knew most of this material once upon a time, but now you need a quick review.

Whatever the reason, I hope that I'm able to give you what you need in order to succeed. Good luck!

Icons Used in This Book

If you've read any other *For Dummies* books (such as the great *Chemistry For Dummies*), you'll recognize the two icons used in this book. Here are their meanings:

This icon alerts you to those really important things you shouldn't forget. These are ideas that you most probably need to memorize for an exam.

This icon points out the easiest or quickest way to understand a particular concept. These are the tricks of the trade that I've picked up in my 40+ years learning chemistry.

Where to Go from Here

Where you go next really depends on you and your reason for using this book. If you're having difficulty with a particular topic, go right to that chapter and section. If you're a real novice, start at Chapter 1 and go from there. If you're using the book for review, skim quickly starting at the beginning and read in more depth those topics that seem a little fuzzy to you. You can even use this book as a fat bookmark in your regular chemistry textbook.

Whatever way you use this book, I hope that it helps and you grow to appreciate the wonderful world of chemistry.

The 5th Wave

By Rich Tennant

"I love this time of year when the organic chemistry students start exploring new and exciting ways for bonding carbon atoms."

Chapter 1

Matter and Energy: Exploring the Stuff of Chemistry

Simply put, chemistry is a whole branch of science about *matter,* which is anything that has mass and occupies space. *Chemistry* is the study of the composition and properties of matter and the changes it undergoes.

Matter and energy are the two basic components of the universe. Scientists used to believe that these two things were separate and distinct, but now they realize that matter and energy are linked. In an atomic bomb or nuclear reactor, for instance, matter is converted into energy. (Perhaps someday science fiction will become a reality and converting the human body into energy and back in a transporter will be commonplace.)

In this chapter, you examine the different states of matter and what happens when matter goes from one state to another. I show you how to use the SI (metric) system to make matter and energy measurements, and I describe types of energy and how energy is measured.

Knowing the States of Matter and Their Changes

Matter is anything that has mass and occupies space. It can exist in one of three classic states: solid, liquid, and gas. When a substance goes from one state of matter to another, the process is called a *change of state,* or *phase change.* Some rather interesting things occur during this process, which I explain in this section.

Solids, liquids, and gases

Particles of matter behave differently depending on whether they're part of a solid, liquid, or gas. As Figure 2-1 shows, the particles may be organized or clumped, close or spread out. In this section, you look at the solid, liquid, and gaseous states of matter.

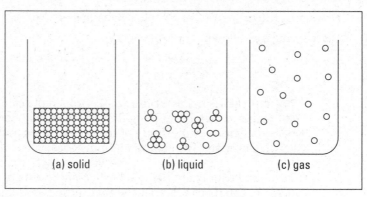

    (a) solid   (b) liquid   (c) gas

Figure 2-1: Solid, liquid, and gaseous states of matter.

Solids

At the *macroscopic level,* the level at which you directly observe with your senses, a solid has a definite shape and occupies a definite volume. Think of an ice cube in a glass — it's a solid. You can easily weigh the ice cube and measure its volume.

At the *microscopic level* (where items are so small that people can't directly observe them), the particles that make up the solid are very close together and aren't moving around very much (see Figure 2-1a). That's because in many solids, the particles are pulled into a rigid, organized structure of repeating patterns called a *crystal lattice*. The particles in the crystal lattice are still moving but barely — it's more of a slight vibration. Depending on the particles, this crystal lattice may be of different shapes.

Liquids

Unlike solids, liquids have no definite shape; however, they do have a definite volume, just like solids do. The particles in liquids are much farther apart than the particles in solids, and they're also moving around much more (see Figure 2-1b).

Even though the particles are farther apart, some particles in liquids may still be near each other, clumped together in small groups. The attractive forces among the particles aren't as strong as they are in solids, which is why liquids don't have a definite shape. However, these attractive forces are strong enough to keep the substance confined in one large mass — a liquid — instead of going all over the place.

Gases

A gas has no definite shape and no definite volume. In a gas, particles are much farther apart than they are in solids or liquids (see Figure 2-1c), and they're moving relatively independent of each other. Because of the distance between the particles and the independent motion of each of them, the gas expands to fill the area that contains it (and thus it has no definite shape).

Condensing and freezing

If you cool a gaseous or liquid substance, you can watch the changes of state, or *phase changes,* that occur. Here are the phase changes that happen as substances lose energy:

- ✓ **Condensation:** When a substance *condenses,* it goes from a gas to a liquid state. Gas particles have a high amount of energy, but as they're cooled, that energy decreases. The attractive forces now have a chance to draw the particles closer together, forming a liquid. The particles are now in clumps, as is characteristic of particles in a liquid state.

✔ **Freezing:** A substance *freezes* when it goes from a liquid to a solid. As energy is removed by cooling, the particles in a liquid start to align themselves, and a solid forms. The temperature at which this occurs is called the *freezing point (fp)* of the substance.

TIP

You can summarize the process of water changing from a gas to a solid in this way:

$$H_2O(g) \rightarrow H_2O(l) \rightarrow H_2O(s)$$

Here, the *(l)* stands for liquid, the *(g)* stands for gas, and *(s)* stands for solid.

Melting and boiling

As a substance heats, it can change from a solid to a liquid to a gas. For water, you represent the change like this:

$$H_2O(s) \rightarrow H_2O(l) \rightarrow H_2O(g)$$

This section explains melting and boiling, the changes of state that occur as a substance gains energy.

From solid to liquid

When a substance melts, it goes from a solid to a liquid state. Here's what happens: If you start with a solid, such as ice, and take temperature readings while heating it, you find that the temperature of the solid begins to rise as the heat causes the particles to vibrate faster and faster in the crystal lattice.

After a while, some of the particles move so fast that they break free of the lattice, and the crystal lattice (which keeps a solid *solid*) eventually breaks apart. The solid begins to go from a solid state to a liquid state — a process called *melting*. The temperature at which melting occurs is called the *melting point (mp)* of the substance. The melting point for ice is 32°F, or 0°C.

REMEMBER

During changes of state, such as melting, the temperature remains constant — even though a liquid contains more energy than a solid. So if you watch the temperature of ice as it melts, you see that the temperature remains steady at 0°C until all the ice has melted.

The melting point (solid to a liquid) is the same as the freezing point (liquid to a solid).

From liquid to gas

The process by which a substance moves from the liquid state to the gaseous state is called *boiling*.

If you heat a liquid, such as a pot of cool water, the temperature of the liquid rises and the particles move faster and faster as they absorb the heat. The temperature rises until the liquid reaches the next change of state — boiling. As the particles heat up and move faster and faster, they begin to break the attractive forces between each other and move freely as a gas, such as steam, the gaseous form of water.

The temperature at which a liquid begins to boil is called the *boiling point (bp)*. The bp depends on atmospheric pressure, but for water at sea level, it's 212°F, or 100°C. The temperature of a boiling substance remains constant until all of it has been converted to a gas.

Skipping liquids: Sublimation

Most substances go through the logical progression from solid to liquid to gas as they're heated (or vice versa as they're cooled). But a few substances go directly from the solid to the gaseous state without ever becoming a liquid. Scientists call this process *sublimation*. Dry ice — solid carbon dioxide, written as $CO_2(s)$ — is the classic example of sublimation. You can see dry ice pieces becoming smaller as the solid begins to turn into a gas, but no liquid forms during this phase change.

The process of sublimation of dry ice is represented as

$$CO_2(s) \rightarrow CO_2(g)$$

Besides dry ice, mothballs and certain solid air fresheners also go through the process of sublimation. The reverse of sublimation is *deposition* — going directly from a gaseous state to a solid state.

Pure Substances and Mixtures

One of the basic processes in science is classification. In this section, I explain how all matter can be classified as either a pure substance or a mixture (see Figure 2-2).

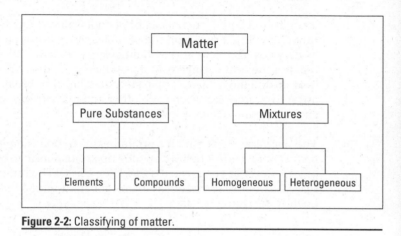

Figure 2-2: Classifying of matter.

Pure substances

A *pure substance,* like salt or sugar, has a definite and constant composition or makeup. A pure substance can be either an element or a compound, but the composition of a pure substance doesn't vary.

Elements

An *element* is composed of a single kind of atom. An *atom* is the smallest particle of an element that still has all the properties of the element. For instance, if you slice and slice a chunk of the element gold until only one tiny particle is left that can't be chopped anymore without losing the properties that make gold *gold,* then you have an atom. (I discuss properties later in the section "Nice Properties You've Got There.")

The atoms in an element all have the same number of protons. *Protons* are subatomic particles — particles of an atom. (Chapter 2 covers the three major subatomic particles in great, gory detail.) The important thing to remember right

now is that elements are the building blocks of matter.
They're represented in the periodic table, which you explore
in Chapter 3.

Compounds

A *compound* is composed of two or more elements in a specific
ratio. For example, water (H_2O) is a compound made up of two
elements, hydrogen (H) and oxygen (O). These elements are
combined in a very specific way — in a ratio of two hydrogen
atoms to one oxygen atom (hence, H_2O). A lot of compounds
contain hydrogen and oxygen, but only one has that special
2-to-1 ratio called *water*.

A compound has physical and chemical properties different
from the elements that make it up. For instance, even though
water is made up of hydrogen and oxygen, water's properties
are a unique combination of the two elements.

Chemists can't easily separate the components of a compound: They have to resort to some type of chemical reaction.

Throwing mixtures into the mix

Mixtures are physical combinations of pure substances that
have no definite or constant composition; the composition of
a mixture varies according to whoever prepares the mixture.
Each component of the mixture retains its own set of physical
and chemical characteristics.

Chemists can easily separate the different parts of a mixture
by physical means, such as filtration. For example, suppose
you have a mixture of salt and sand, and you want to purify
the sand by removing the salt. You can do this by adding
water, dissolving the salt, and then filtering the mixture. You
then end up with pure sand.

Mixtures can be either homogeneous or heterogeneous:

> ✔ **Homogeneous mixtures:** Sometimes called *solutions,*
> homogeneous mixtures are relatively uniform in compo-
> sition. Every portion of the mixture is like every other
> portion. If you dissolve sugar in water and mix it really
> well, your mixture is basically the same no matter where
> you sample it. I cover solutions in Chapter 10.

✔ **Heterogeneous mixtures:** The composition of heterogeneous mixtures varies from position to position within the sample. For instance, if you put some sugar in a jar, add some sand, and then give the jar a couple of shakes, your mixture doesn't have the same composition throughout the jar. Because the sand is heavier, there's probably more sand at the bottom of the jar and more sugar at the top.

Measuring Matter

Scientists often make measurements, which may include such things as mass, volume, and temperature. If each nation had its own measurement system, communication among scientists would be tremendously hampered, so scientists adopted a worldwide measurement system to ensure they can speak the same language.

The *SI system* (from the French *Système international*) is a worldwide measurement system based on the older metric system. SI is a decimal system with basic units for things like mass, length, and volume and prefixes that modify the basic units. For example, here are some very useful SI prefixes:

✔ *kilo- (k)* means 1,000

✔ *centi- (c)* means 0.01

✔ *milli- (m)* means 0.001

So a kilogram (kg) is 1,000 grams, and a kilometer (km) is 1,000 meters. A milligram (mg) is 0.001 grams — or you can say that there are 1,000 milligrams in a gram.

Here are some basic SI units and how they compare to the English units common in the U.S.:

✔ **Length:** The basic unit of length in the SI system is the *meter (m)*. A meter is a little longer than a yard; 1.094 yards are in a meter. The most useful SI/English conversion for length is 2.54 centimeters = 1 inch

✔ **Mass:** The basic unit of mass in the SI system for chemists is the *gram (g)*. And the most useful conversion for mass is 454 grams = 1 pound

✔ **Volume:** The basic unit for volume in the SI system is the *liter (L)*. The most useful conversion is 0.946 liter = 1 quart

Suppose you want to find the weight of a 5.0-lb. bag of potatoes in kilograms. The setup would look that this:

$$\frac{5.0 \ \text{lbs}}{1} \cdot \frac{454 \ \text{g}}{1 \ \text{lb}} \cdot \frac{1 \ \text{kg}}{1000 \ \text{g}} = 2.3 \ \text{kg}$$

Nice Properties You've Got There

When chemists study chemical substances, they examine two types of properties:

✔ **Chemical properties:** These properties enable a substance to change into a brand-new substance, and they describe how a substance reacts with other substances. Does a substance change into something completely new when water is added — like sodium metal changes to sodium hydroxide? Does the substance burn in air?

✔ **Physical properties:** These properties describe the physical characteristics of a substance. The mass, volume, and color of a substance are physical properties, and so is its ability to conduct electricity. Physical properties can be extensive or intensive:

- *Extensive properties,* such as mass and volume, depend on the amount of matter present.

- *Intensive properties,* such as color and density, don't depend on the amount of matter present. A large chunk of gold, for example, is the same color as a small chunk of gold.

Intensive properties are especially useful to chemists because intensive properties can be used to identify a substance. For example, knowing the differences between the density of quartz and diamond allows a jeweler to check out that engagement ring quickly and easily.

Density (d) is the ratio of the mass *(m)* to volume *(v)* of a substance. Mathematically, it looks like this:

$$d = m/v$$

Usually, mass is described in grams (g) and volume is described in milliliters (mL), so density is g/mL. Because the volumes of liquids vary somewhat with temperature, chemists usually specify the temperature at which they made a density measurement. Most reference books report densities at 20°C, because it's close to room temperature and easy to measure without a lot of heating or cooling. The density of water at 20°C, for example, is 1 g/mL.

 You may sometimes see density reported as g/cm³ or g/cc, both of which mean *grams per cubic centimeter*. These units are the same as g/mL.

Calculating density is pretty straightforward. You measure the mass of an object by using a balance or scale, determine the object's volume, and then divide the mass by the volume.

 With an irregular solid, like a rock, you can measure the volume by using the Archimedes principle. The *Archimedes principle* states that the volume of a solid is equal to the volume of water it displaces. Simply read the volume of water in a container, submerge the solid object, and read the volume level again. The difference is the volume of the object.

Energy Types

Matter is one of two components of the universe. Energy is the other. *Energy* is the ability to do work.

Energy can take several forms, such as heat energy, light energy, electrical energy, and mechanical energy. But two general categories of energy are especially important to chemists: kinetic energy and potential energy.

Kinetic energy

Kinetic energy is energy of motion. A baseball flying through the air toward a batter has a large amount of kinetic energy — just ask anyone who's ever been hit with a baseball.

Chemists sometimes study moving particles, especially gases, because the kinetic energy of these particles helps determine whether a particular reaction may take place. As particles collide, kinetic energy may be transferred from one particle to another, causing chemical reactions.

Kinetic energy can be converted into other types of energy. In a hydroelectric dam, the kinetic energy of the falling water is converted into electrical energy. In fact, a scientific law — *the law of conservation of energy* — states that in ordinary chemical reactions (or physical processes), energy is neither created nor destroyed, but it can be converted from one form to another.

Potential energy

Potential energy is stored energy. Objects may have potential energy stored in terms of their position. A ball up in a tree has potential energy due to its height. If that ball were to fall, that potential energy would be converted to kinetic energy.

Potential energy due to position isn't the only type of potential energy. Chemists are far more interested in the energy stored (potential energy) in *chemical bonds,* which are the forces that hold atoms together in compounds.

Human bodies store energy in chemical bonds. When you need that energy, your body can break those bonds and release it. The same is true of the fuels people commonly use to heat their homes and run their automobiles. Energy is stored in these fuels — gasoline, for example — and is released when chemical reactions take place.

Temperature and Heat

When you measure, say, the air temperature in your backyard, you're really measuring the average *kinetic energy* (the energy of motion) of the gas particles in your backyard. The faster those particles are moving, the higher the temperature is.

The temperature reading from your thermometer is related to the *average* kinetic energy of the particles. Not all the particles are moving at the same speed. Some are going very fast, and some are going relatively slow, but most are moving at a speed between the two extremes.

If you're in the U.S., you probably use the Fahrenheit scale to measure temperatures, but most scientists use either the Celsius (°C) or Kelvin (K) temperature scale. (***Remember:*** There's no degree symbol associated with K.) Water boils at 100°C (373 K) and freezes at 0°C (273 K).

Here's how to do some temperature conversions:

- ✔ **Fahrenheit to Celsius:** °C = ⅚(°F − 32)

- ✔ **Celsius to Fahrenheit:** °F = ⅚(°C) + 32

- ✔ **Celsius to Kelvin:** K = °C + 273

Heat is not the same as temperature. When you measure the *temperature* of something, you're measuring the average kinetic energy of the individual particles. *Heat,* on the other hand, is the amount of energy that goes from one substance to another.

The unit of heat in the SI system is the *joule (J)*. Most people still use the metric unit of heat, the *calorie (cal)*. Here's the relationship between the two:

   1 calorie = 4.184 joules

The calorie is a fairly small amount of heat: the amount it takes to raise the temperature of 1 gram of water 1°C. I often use the *kilocalorie (kcal),* which is 1,000 calories, as a convenient unit of heat. If you burn a large kitchen match completely, it produces about 1 kcal.

Chapter 2

What's In an Atom?

*I*n this chapter, I tell you about atoms, the fundamental building blocks of the universe. I cover the three basic particles of an atom — protons, neutrons, and electrons — and show you where they're located. And I spend quite a bit of time discussing electrons themselves, because chemical reactions (where a lot of chemistry comes into play) depend on the loss, gain, or sharing of electrons.

Subatomic Particles

The *atom* is the smallest part of matter that represents a particular element. For quite a while, the atom was thought to be the smallest part of matter that could exist. But in the latter part of the 19th century and early part of the 20th, scientists discovered that atoms are composed of certain subatomic particles and that no matter what the element, the same subatomic particles make up the atom. The number of the various subatomic particles is the only thing that varies.

Scientists now recognize that there are many subatomic particles (this really makes physicists salivate). But to be successful in chemistry, you really only need to be concerned with the three major subatomic particles:

- Protons

- Neutrons

- Electrons

Table 2-1 summarizes the characteristics of these three subatomic particles. The masses of the subatomic particles are listed in two ways: grams and *amu,* which stands for *atomic mass units.* Expressing mass in amu is much easier than using the gram equivalent.

Table 2-1   The Three Major Subatomic Particles

Name	Symbol	Charge	Mass (g)	Mass (amu)	Location
Proton	p^+	+1	1.673×10^{-24}	1	In the nucleus
Neutron	n^0	0	1.675×10^{-24}	1	In the nucleus
Electron	e^-	−1	9.109×10^{-28}	0.0005	Outside the nucleus

Atomic mass units are based on something called the *carbon-12 scale,* a worldwide standard that's been adopted for atomic weights. By international agreement, a carbon atom that contains six protons and six neutrons has an atomic weight of exactly 12 amu, so *1 amu* is defined as $\frac{1}{12}$ of this carbon atom. Because the masses in grams of protons and neutrons are almost exactly the same, both protons and neutrons are said to have a mass of 1 amu. Notice that the mass of an electron is much smaller than that of either a proton or neutron. It takes almost 2,000 electrons to equal the mass of a single proton.

Table 2-1 also shows the electrical charge associated with each subatomic particle. Matter can be electrically charged in one of two ways: positive or negative. The proton carries one unit of positive charge, the electron carries one unit of negative charge, and the neutron has no charge — it's neutral.

Scientists have discovered through observation that objects with like charges, whether positive or negative, repel each other, and objects with unlike charges attract each other.

The atom itself has no charge. It's neutral. (Well, actually, certain atoms can gain or lose electrons and acquire a charge, as I explain in the later section "Ions: Varying electrons." Atoms that gain a charge, either positive or negative, are called *ions.*) So how can an atom be neutral if it contains positively charged protons and negatively charged electrons? The answer is that there are *equal* numbers of protons and electrons — equal numbers of positive and negative charges — so they cancel each other out.

The last column in Table 2-1 lists the location of the three subatomic particles. Protons and neutrons are located in the *nucleus,* a dense central core in the middle of the atom, and the electrons are located outside the nucleus (for details, see "Locating Those Electrons?" later in this chapter).

Centering on the Nucleus

In 1911, Ernest Rutherford discovered that atoms have a nucleus — a center — containing protons. Scientists later discovered that the nucleus also houses the neutron.

The nucleus is very, very small and very, very dense when compared to the rest of the atom. Typically, atoms have diameters that measure around 10^{-10} meters (that's small!). Nuclei are around 10^{-15} meters in diameter (that's *really* small!). If the Superdome in New Orleans represented a hydrogen atom, the nucleus would be about the size of a pea.

The protons of an atom are all crammed together inside the nucleus. Now you may be thinking, "Okay, each proton carries a positive charge, and like charges repel each other. So if all the protons are repelling each other, why doesn't the nucleus simply fly apart?" It's the Force, Luke. Forces in the nucleus counteract this repulsion and hold the nucleus together. Physicists call these forces *nuclear glue.* (*Note:* Sometimes this "glue" isn't strong enough, and the nucleus does break apart. This process is called *radioactivity,* and I cover it in Chapter 4.)

Not only is the nucleus very small, but it also contains most of the mass of the atom. In fact, for all practical purposes, the mass of the atom is the sum of the masses of the protons and neutrons. (I ignore the minute mass of the electrons unless I'm doing very, very precise calculations.)

The sum of the number of protons plus the number of neutrons in an atom is called the *mass number*. And the number of protons in a particular atom is given a special name, the *atomic number*. Chemists commonly use the symbolization in Figure 2-1 to represent these amounts for a particular element.

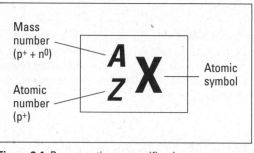

Mass number ($p^+ + n^0$)

Atomic number (p^+)

$$^A_Z X$$

Atomic symbol

Figure 2-1: Representing a specific element.

As Figure 2-1 shows, chemists use the placeholder X to represent the chemical symbol. You can find an element's chemical symbol on the periodic table or in a list of elements. The placeholder Z represents the atomic number — the number of protons in the nucleus. And A represents the mass number, the sum of the number of protons plus neutrons. The mass number is listed in amu.

For example, you can represent a uranium atom that has 92 protons and a mass number of 238 as in Figure 2-2.

$$^{238}_{92} U$$

Figure 2-2: Representing uranium.

You can find the number of neutrons in an atom by subtracting the atomic number (number of protons) from the mass number (protons plus neutrons). For instance, you know that uranium has an atomic number of 92 and mass number of 238.

So if you want to know the number of neutrons in uranium, all you have to do is subtract the atomic number (92 protons) from the mass number (238 protons plus neutrons). The answer shows that uranium has 146 neutrons.

But how many electrons does uranium have? Because the atom is neutral (it has no electrical charge), there must be equal numbers of positive and negative charges inside it, or equal numbers of protons and electrons. So there are 92 electrons in each uranium atom.

You can find both the element symbol and its atomic number on the periodic table, but the mass number for a particular element is not shown there. What is shown is the average *atomic mass* or *atomic weight* for all forms of that particular element, taking into account the percentages of each found in nature. See the later section "Isotopes: Varying neutrons" for details on other forms of an element.

Locating Those Electrons

Many of the important topics in chemistry, such as chemical bonding, the shape of molecules, and so on, are based on where the electrons in an atom are located. Simply saying that the electrons are located outside the nucleus isn't good enough; chemists need to have a much better idea of their location, so this section helps you figure out where you can find those pesky electrons.

The quantum mechanical model

Early models of the atom had electrons going around the nucleus in a random fashion. But as scientists discovered more about the atom, they found that this representation probably wasn't accurate. Today, scientists use the quantum mechanical model, a highly mathematical model, to represent the structure of the atom.

This model is based on *quantum theory,* which says that matter also has properties associated with waves. According to quantum theory, it's impossible to know an electron's exact position and *momentum* (speed and direction, multiplied by mass) at the same time. This is known as the *uncertainty principle.* So scientists had to develop the concept of *orbitals* (sometimes called *electron clouds*), volumes of space in which an electron is likely present. In other words, certainty was replaced with probability.

The quantum mechanical model of the atom uses complex shapes of orbitals. Without resorting to a lot of math (you're welcome), this section shows you some aspects of this newest model of the atom.

Scientists introduced four numbers, called *quantum numbers,* to describe the characteristics of electrons and their orbitals. You'll notice that they were named by top-rate techno-geeks:

- ✔ Principal quantum number n

- ✔ Angular momentum quantum number l

- ✔ Magnetic quantum number m_l

- ✔ Spin quantum number m_s

Table 2-2 summarizes the four quantum numbers. When they're all put together, theoretical chemists have a pretty good description of the characteristics of a particular electron.

Table 2-2   Summary of the Quantum Numbers

Name	Symbol	Description	Allowed Values
Principal	n	Orbital energy	Positive integers (1, 2, 3, and so on)
Angular momentum	l	Orbital shape	Integers from 0 to $n-1$
Magnetic	m_l	Orientation	Integers from $-l$ to $+l$
Spin	m_s	Electron spin	$+\frac{1}{2}$ or $-\frac{1}{2}$

The principal quantum number n

The principal quantum number n describes the average distance of the orbital from the nucleus — and the energy of the electron in an atom. It can have only positive integer (whole-number) values: 1, 2, 3, 4, and so on. The larger the value of n, the higher the energy and the larger the orbital, or electron shell.

The angular momentum quantum number l

The angular momentum quantum number l describes the shape of the orbital, and the shape is limited by the principal quantum number n. The angular momentum quantum number l can have positive integer values from 0 to $n-1$. For example, if the n value is 3, three values are allowed for l: 0, 1, and 2.

The value of l defines the shape of the orbital, and the value of n defines the size.

Orbitals that have the same value of n but different values of l are called *subshells*. These subshells are given different letters to help chemists distinguish them from each other. Table 2-3 shows the letters corresponding to the different values of l.

Table 2-3	Letter Designation of the Subshells
Value of l (Subshell)	*Letter*
0	s
1	p
2	d
3	f
4	g

When chemists describe one particular subshell in an atom, they can use both the n value and the subshell letter — 2p, 3d, and so on. Normally, a subshell value of 4 is the largest needed to describe a particular subshell. If chemists ever need a larger value, they can create subshell numbers and letters.

Figure 2-3 shows the shapes of the s, p, and d orbitals. In Figure 2-3a, there are two s orbitals — one for energy level 1 (1s) and the other for energy level 2 (2s). S orbitals are spherical with the nucleus at the center. Notice that the 2s orbital is larger in diameter than the 1s orbital. In large atoms, the 1s orbital is nestled inside the 2s, just like the 2p is nestled inside the 3p.

Figure 2-3b shows the shapes of the p orbitals, and Figure 2-3c shows the shapes of the d orbitals. Notice that the shapes get progressively more complex.

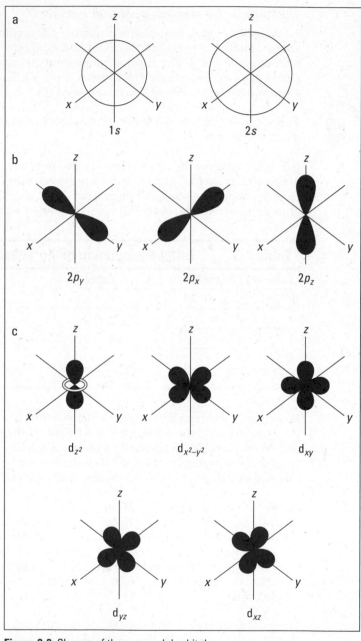

Figure 2-3: Shapes of the s, p, and d orbitals.

The magnetic quantum number m_l

The magnetic quantum number m_l describes how the various orbitals are oriented in space. The value of m_l depends on the value of l. The values allowed are integers from $-l$ to 0 to $+l$. For example, if the value of l = 1 (p orbital — see Table 3-4), you can write three values for m_l: –1, 0, and +1. This means that there are three different p subshells for a particular orbital. The subshells have the same energy but different orientations in space.

Figure 2-3b shows how the p orbitals are oriented in space. Notice that the three p orbitals correspond to m_l values of –1, 0, and +1, oriented along the x, y, and z axes.

The spin quantum number m_s

The fourth and final quantum number is the spin quantum number m_s. This one describes the direction the electron is spinning in a magnetic field — either clockwise or counter-clockwise. Only two values are allowed for m_s: +½ or –½. For each subshell, there can be only two electrons, one with a spin of +½ and another with a spin of –½.

Putting the quantum numbers together

Table 2-4 summarizes the quantum numbers available for the first two energy levels.

Table 2-4		Quantum Numbers for the First Two Energy Levels		
n	l	Subshell Notation	m_l	m_s
1	0	1s	0	+½, –½
2	0	2s	0	+½, –½
	1	2p	–1	+½, –½
			0	+½, –½
			+1	+½, –½

Table 2-4 shows that in energy level 1 (n = 1), there's only an s orbital. There's no p orbital because an l value of 1 (p orbital) is not allowed. And notice that there can be only two electrons in that 1s orbital (m_s of +½ and –½). In fact, there can be only two electrons in any s orbital, whether it's 1s or 5s.

Each time you move higher in a major energy level, you add another orbital type. So when you move from energy level 1 to energy level 2 ($n = 2$), there can be both s and p orbitals. If you write out the quantum numbers for energy level 3, you see s, p, and d orbitals.

Notice also that there are three subshells (m_l) for the 2p orbital (see Figure 2-3b) and that each holds a maximum of two electrons. The three 2p subshells can hold a maximum of six electrons.

There's an energy difference in the major energy levels (energy level 2 is higher in energy than energy level 1), but there's also a difference in the energies of the different orbitals within an energy level. At energy level 2, both s and p orbitals are present. But the 2s is lower in energy than the 2p. The three subshells of the 2p orbital have the same energy. Likewise, the five subshells of the d orbitals (see Figure 2-3c) have the same energy.

Energy level diagrams

Chemists find quantum numbers useful when they're looking at chemical reactions and bonding (and those are things many chemists like to study). But they find two other representations for electrons — energy level diagrams and electron configurations — more useful and easier to work with.

Chemists use both of these things to represent which energy level, subshell, and orbital are occupied by electrons in any particular atom. Chemists use this information to predict what type of bonding will occur with a particular element and to show exactly which electrons are being used. These representations are also useful in showing why certain elements behave in similar ways.

In this section, I show you how to use an energy level diagram and write electron configurations. I also discuss valence electrons, which are key in chemical reactions.

The dreaded energy level diagram

Figure 2-4 is a blank energy level diagram you can use to depict electrons for any particular atom. It doesn't show all the known orbitals and subshells, but with this diagram, you should be able to do most anything you need to.

I represent orbitals with dashes in which you can place a maximum of two electrons. The 1s orbital is closest to the nucleus, and it has the lowest energy. It's also the only orbital in energy level 1 (refer to Table 2-4). At energy level 2, there are both s and p orbitals, with the 2s having lower energy than the 2p. The three 2p subshells are represented by three dashes of the same energy. The figure also shows energy levels 3, 4, and 5.

Notice that the 4s orbital has lower energy than the 3d: This is an exception to what you may have thought, but it's what's observed in nature. Go figure.

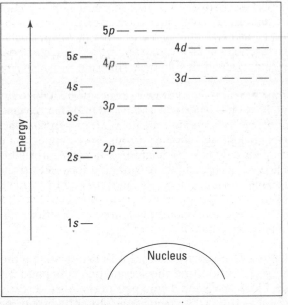

Figure 2-4: An energy level diagram.

Speaking of which, Figure 2-5 shows the *Aufbau principle*, a method for remembering the order in which orbitals fill the vacant energy levels.

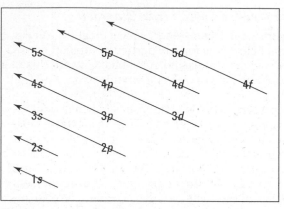

Figure 2-5: The Aufbau filling chart.

In using the energy level diagram, remember two things:

✔ Electrons fill the lowest vacant energy levels first.

✔ When there's more than one subshell at a particular energy level, such as at the 3p or 4d levels (see Figure 2-4), only one electron fills each subshell until each subshell has one electron. Then electrons start pairing up in each subshell. This rule is named *Hund's rule.*

Suppose you want to draw the energy level diagram of oxygen. You look on the periodic table and find that oxygen is atomic number 8. This number means that oxygen has eight protons in its nucleus and eight electrons. So you put eight electrons into your energy level diagram. You can represent electrons as arrows, as in Figure 2-6. Note that if two electrons end up in the same orbital, one arrow faces up and the other faces down. This is called *spin pairing.* It corresponds to the +½ and –½ of m_s (see "The spin quantum number m_s" section, earlier in this chapter, for details).

The first electron goes into the 1s orbital, filling the lowest energy level first, and the second one spin-pairs with the first one. Electrons 3 and 4 spin-pair in the next-lowest vacant orbital — the 2s. Electron 5 goes into one of the 2p subshells (no, it doesn't matter which one — they all have the same energy), and electrons 6 and 7 go into the other two totally vacant 2p orbitals. The last electron spin-pairs with one of the electrons in the 2p subshells (again, it doesn't matter which one you pair it with). Figure 2-6 shows the completed energy level diagram for oxygen.

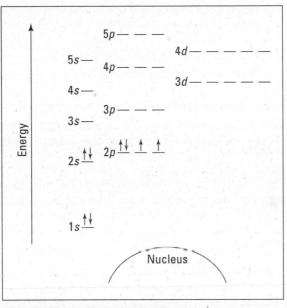

Figure 2-6: Energy level diagram for oxygen.

Electron configurations

Energy level diagrams are useful when you need to figure out chemical reactions and bonding, but they're very bulky to work with. Wouldn't it be nice if there were another representation that gives just about the same information but in a much more concise form? Well, there is. It's called the *electron configuration*.

The electron configuration for oxygen is $1s^2 2s^2 2p^4$. Compare that notation with the energy level diagram for oxygen in Figure 2-6. Doesn't the electron configuration take up a lot less space? You can derive the electron configuration from the energy level diagram. The first two electrons in oxygen fill the 1s orbital, so you show it as $1s^2$ in the electron configuration. The 1 is the energy level, the s represents the type of orbital, and the superscript 2 represents the number of electrons in that orbital. The next two electrons are in the 2s orbital, so you write $2s^2$. And finally, you show the four electrons in the 2p orbital as $2p^4$. Put it all together, and you get $1s^2 2s^2 2p^4$.

The sum of the superscript numbers equals the atomic number, or the number of electrons in the atom.

Here are a couple of electron configurations you can use to check your conversions from energy level diagrams:

Chlorine (Cl): $1s^2 2s^2 2p^6 3s^2 3p^5$

Iron (Fe): $1s^2 2s^2 2p^6 3s^2 3p^6 4s^2 3d^6$

Valence electrons: Clues about chemical reactions

Knowing the number of electrons that are an atom's outer-most energy level gives you a big clue about how that atom will react.

When chemists study chemical reactions, they study the transfer or sharing of electrons. The electrons more loosely held by the nucleus — the electrons in the energy level far-thest away from the nucleus — are the ones that are gained, lost, or shared.

Electrons are negatively charged, and the nucleus has a positive charge due to the protons. The protons attract and hold the elec-trons, but the farther away the electrons are, the less the attrac-tive force.

The electrons in the outermost energy level are commonly called *valence electrons*. Chemists really only consider the electrons in the s and p orbitals in the energy level that's cur-rently being filled as valence electrons. In the electron config-uration for oxygen, $1s^2 2s^2 2p^4$, energy level 1 is filled, and there are two electrons in the 2s orbital and four electrons in the 2p orbital for a total of six valence electrons. Those valence elec-trons are the ones lost, gained, or shared.

Isotopes and Ions

The number of protons in an atom determines which element you have. But sometimes the number of neutrons or electrons varies, so you see several different versions of the atoms of that element. In this section, I introduce you to two variations: isotopes and ions.

Isotopes: Varying neutrons

The atoms of a particular element can have an identical number of protons and electrons but varying numbers of neutrons. If they have different numbers of neutrons, then the atoms are called *isotopes*.

Hydrogen is a common element here on Earth. Hydrogen's atomic number is 1 — its nucleus contains 1 proton. The hydrogen atom also has 1 electron. Because it has the same number of protons as electrons, the hydrogen atom is neutral (the positive and negative charges have canceled each other out).

Most of the hydrogen atoms on Earth contain no neutrons. You can use the symbolization in Figure 2-1 to represent hydrogen atoms that don't contain neutrons, as shown Figure 2-7a shows.

However, approximately one hydrogen atom out of 6,000 contains a neutron in its nucleus. These atoms are still hydrogen, because they each have one proton; they simply have a neutron as well, which most hydrogen atoms lack. So these atoms are called isotopes. Figure 2-7b shows an isotope of hydrogen, commonly called *deuterium*. It's still hydrogen, because it contains only one proton, but it's different from the hydrogen in Figure 2-7a, because it also has one neutron. Because it contains one proton and one neutron, its mass number is 2 amu.

There's even an isotope of hydrogen containing two neutrons. This one's called *tritium*, and it's represented in Figure 2-7c. Tritium is extremely rare, but it can easily be created.

Figure 2-7 also shows an alternative way of representing isotopes: Write the element symbol, a dash, and then the mass number.

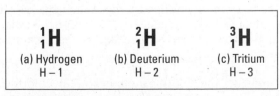

^1_1H	^2_1H	^3_1H
(a) Hydrogen	(b) Deuterium	(c) Tritium
H – 1	H – 2	H – 3

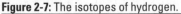

Figure 2-7: The isotopes of hydrogen.

Now you may be wondering, "If I'm doing a calculation involving atomic mass, which isotope do I use?" Well, you use an average of all the naturally occurring isotopes of that element — but not a simple average. Instead, you use a *weighted average,* which takes into consideration the abundances of the naturally occurring isotopes. You find this number on the periodic table.

For hydrogen, you have to take into consideration that there's a *lot* more H-1 than H-2 and only a very tiny amount of H-3. That's why the atomic mass of hydrogen on the periodic table isn't a whole number: It's 1.0079 amu. The number shows that there's a lot more H-1 than H-2 and H-3.

Ions: Varying electrons

Because an atom itself is neutral, I say that the number of protons and electrons in atoms are equal throughout this book. But in some cases, an atom can acquire an electrical charge. For example, in the compound sodium chloride — table salt — the sodium atom has a positive charge and the chlorine atom has a negative charge. Atoms (or groups of atoms) in which there are unequal numbers of protons and electrons are called *ions.*

The neutral sodium atom has 11 protons and 11 electrons, which means it has 11 positive charges and 11 negative charges. Overall, the sodium atom is neutral, and it's represented like this: Na. But the sodium *ion* contains one more positive charge than negative charge, so it's represented like this: Na^+ (the $^+$ represents its net positive electrical charge).

Gaining and losing electrons

Atoms become ions by gaining or losing electrons. And ions that have a positive charge are called *cations.* The progression goes like this: The Na^+ ion is formed from the loss of one electron. Because it lost an electron, it has more protons than electrons, or more *positive* charges than negative charges, which means it's now called the Na^+ cation. Likewise, the Mg^{2+} cation is formed when the neutral magnesium atom loses two electrons.

Now consider the chlorine atom in sodium chloride. The neutral chlorine atom has acquired a negative charge by gaining an electron. Because it has unequal numbers of protons and electrons, it's now an ion, represented like this: Cl^-. And because ions that have a negative charge are called *anions*, it's now called the Cl^- anion. (You can get the full scoop on ions, cations, and anions in Chapter 5, if you're interested. This here's just a teaser.)

Writing electron configurations

Here are some extra tidbits about ions for your chemistry reading pleasure:

- You can write electron configurations and energy level diagrams for ions. The neutral sodium atom (11 protons) has an electron configuration of $1s^2 2s^2 2p^6 3s^1$. The sodium cation has lost an electron — the valence electron, which is *farthest* away from the nucleus (the 3s electron, in this case). The electron configuration of Na^+ is $1s^2 2s^2 2p^6$.

- The electron configuration of the chloride ion (Cl^-) is $1s^2 2s^2 2p^6 3s^2 3p^6$. This is the same electron configuration as the neutral argon atom. If two chemical species have the same electron configuration, they're said to be *isoelectronic*. Figuring out chemistry requires learning a whole new language, eh?

- This section has been discussing *monoatomic* (one atom) ions. But *polyatomic* (many atom) ions do exist. The ammonium ion, NH_4^+, is a polyatomic ion, or specifically, a *polyatomic cation*. The nitrate ion, NO_3^-, is also a polyatomic ion, or specifically, a *polyatomic anion*.

Predicting types of bonds

Ions are commonly found in a class of compounds called *salts*, or *ionic solids*. Salts, when melted or dissolved in water, yield solutions that conduct electricity. A substance that conducts electricity when melted or dissolved in water is called an *electrolyte*. Table salt — sodium chloride — is a good example.

On the other hand, when table sugar (sucrose) is dissolved in water, it becomes a solution that doesn't conduct electricity. So sucrose is a *nonelectrolyte*.

Whether a substance is an electrolyte or a nonelectrolyte gives clues to the type of bonding in the compound. If the substance is an electrolyte, the compound is probably *ionically bonded* (see Chapter 5). If it's a nonelectrolyte, it's probably *covalently bonded* (see Chapter 6).

Chapter 3

The Periodic Table

- -

- -

*C*hemists like to put things together into groups based on similar properties. This process, called *classification,* makes studying a particular system much easier. Scientists grouped the elements in the periodic table so they don't have to memorize the properties of individual elements. With the periodic table, they can just remember the properties of the various groups.

The periodic table is the most important tool a chemist possesses. So in this chapter, I show you how the elements are arranged in the table, and I show you some important groups. I also explain how chemists and other scientists use the periodic table.

Repeating Patterns: The Modern Periodic Table

In nature, as well as in things that humankind invents, you may notice some repeating patterns. The seasons repeat their pattern of fall, winter, spring, and summer. The tides repeat their pattern of rising and falling. Tuesday follows Monday, December follows November, and so on. A pattern of repeating order is called *periodicity*.

In the mid-1800s, Dmitri Mendeleev, a Russian chemist, noticed a repeating pattern of chemical properties in the elements that were known at the time. Mendeleev arranged the elements in order of increasing atomic mass (see Chapter 2 for a description of atomic mass) to form something that fairly closely resembles the modern periodic table. He was even able to predict the properties of some of the then-unknown elements. Later, the elements were rearranged in order of increasing *atomic number,* the number of protons in the nucleus of the atom. Figure 3-1 shows the modern periodic table.

PERIODIC TABLE OF THE ELEMENTS

	1 IA	2 IIA	3 IIIB	4 IVB	5 VB	6 VIB	7 VIIB	8 VIIIB	9 VIIIB
1	1 H Hydrogen 1.00797								
2	3 Li Lithium 6.939	4 Be Beryllium 9.0122							
3	11 Na Sodium 22.9898	12 Mg Magnesium 24.312							
4	19 K Potassium 39.102	20 Ca Calcium 40.08	21 Sc Scandium 44.956	22 Ti Titanium 47.90	23 V Vanadium 50.942	24 Cr Chromium 51.996	25 Mn Manganese 54.9380	26 Fe Iron 55.847	27 Co Cobalt 58.9332
5	37 Rb Rubidium 85.47	38 Sr Strontium 87.62	39 Y Yttrium 88.905	40 Zr Zirconium 91.22	41 Nb Niobium 92.906	42 Mo Molybdenum 95.94	43 Tc Technetium (99)	44 Ru Ruthenium 101.07	45 Rh Rhodium 102.905
6	55 Cs Cesium 132.905	56 Ba Barium 137.34	57 La Lanthanum 138.91	72 Hf Hafnium 179.49	73 Ta Tantalum 180.948	74 W Tungsten 183.85	75 Re Rhenium 186.2	76 Os Osmium 190.2	77 Ir Iridium 192.2
7	87 Fr Francium (223)	88 Ra Radium (226)	89 Ac Actinium (227)	104 Rf Rutherfordium (261)	105 Db Dubnium (262)	106 Sg Seaborgium (266)	107 Bh Bohrium (264)	108 Hs Hassium (269)	109 Mt Meitnerium (268)

Lanthanide Series	58 Ce Cerium 140.12	59 Pr Praseodymium 140.907	60 Nd Neodymium 144.24	61 Pm Promethium (145)	62 Sm Samarium 150.35	63 Eu Europium 151.96
Actinide Series	90 Th Thorium 232.038	91 Pa Protactinium (231)	92 U Uranium 238.03	93 Np Neptunium (237)	94 Pu Plutonium (242)	95 Am Americium (243)

								18 VIIIA
			13 IIIA	14 IVA	15 VA	16 VIA	17 VIIA	2 He Helium 4.0026
			5 B Boron 10.811	6 C Carbon 12.01115	7 N Nitrogen 14.0067	8 O Oxygen 15.9994	9 F Fluorine 18.9984	10 Ne Neon 20.183
10 VIIIB	11 IB	12 IIB	13 Al Aluminum 26.9815	14 Si Silicon 28.086	15 P Phosphorus 30.9738	16 S Sulfur 32.064	17 Cl Chlorine 35.453	18 Ar Argon 39.948
28 Ni Nickel 58.71	29 Cu Copper 63.546	30 Zn Zinc 65.37	31 Ga Gallium 69.72	32 Ge Germanium 72.59	33 As Arsenic 74.9216	34 Se Selenium 78.96	35 Br Bromine 79.904	36 Kr Krypton 83.80
46 Pd Palladium 106.4	47 Ag Silver 107.868	48 Cd Cadmium 112.40	49 In Indium 114.82	50 Sn Tin 118.69	51 Sb Antimony 121.75	52 Te Tellurium 127.60	53 I Iodine 126.9044	54 Xe Xenon 131.30
78 Pt Platinum 195.09	79 Au Gold 196.967	80 Hg Mercury 200.59	81 Tl Thallium 204.37	82 Pb Lead 207.19	83 Bi Bismuth 208.980	84 Po Polonium (210)	85 At Astatine (210)	86 Rn Radon (222)
110 Uun Ununnilium (269)	111 Uuu Unununium (272)	112 Uub Ununbium (277)	113 Uut §	114 Uuq Ununquadium (285)	115 Uup §	116 Uuh Ununhexium (289)	117 Uus §	118 Uuo Ununoctium (293)

64 Gd Gadolinium 157.25	65 Tb Terbium 158.924	66 Dy Dysprosium 162.50	67 Ho Holmium 164.930	68 Er Erbium 167.26	69 Tm Thulium 168.934	70 Yb Ytterbium 173.04	71 Lu Lutetium 174.97
96 Cm Curium (247)	97 Bk Berkelium (247)	98 Cf Californium (251)	99 Es Einsteinium (254)	100 Fm Fermium (257)	101 Md Mendelevium (258)	102 No Nobelium (259)	103 Lr Lawrencium (260)

§ Note: Elements 113, 115, and 117 are not known at this time, but are included in the table to show their expected positions.

Figure 3-1: The periodic table.

Arranging Elements in the Periodic Table

Look at the periodic table in Figure 3-1. The elements are arranged in order of increasing atomic number. The *atomic number* (number of protons) is located right above the element symbol. Under the element symbol is the atomic mass, or atomic weight. *Atomic mass* is a weighted average of all naturally occurring isotopes (see Chapter 2 for details).

Notice that two rows of elements — Ce–Lu (commonly called the *lanthanides*) and Th–Lr (the *actinides*) — have been pulled out of the main body of the periodic table. If they were included in the main body of the periodic table, the table would be much wider.

Using the periodic table, you can classify the elements in many ways. Two quite useful ways are

- ✔ Metals, nonmetals, and metalloids
- ✔ Families and periods

Grouping metals, nonmetals, and metalloids

Elements can be metals, nonmetals, or metalloids. In this section, I explain their properties.

Metals

If you look carefully at Figure 3-1, you can see a stair-stepped line starting at boron (B), atomic number 5, and going all the way down to polonium (Po), atomic number 84. Except for germanium (Ge) and antimony (Sb), all the elements to the left of that line can be classified as *metals*. Figure 3-2 shows the metals.

These metals have properties that you normally associate with the metals you encounter in everyday life. They're solid at room temperature (with the exception of mercury, Hg, a liquid), shiny, good conductors of electricity and heat, *ductile* (they can be drawn into thin wires), and *malleable* (they can be easily hammered into very thin sheets). All these metals tend to lose electrons easily (see Chapter 5 for more info). As you can see, the vast majority of the elements on the periodic table are classified as metals.

IA	IIA (2)	IIIB (3)	IVB (4)	VB (5)	VIB (6)	VIIB (7)	VIIIB (8)	VIIIB (9)	VIIIB (10)	IB (11)	IIB (12)	IIIA (13)
3 Li Lithium 6.939	4 Be Beryllium 9.0122											
11 Na Sodium 22.9898	12 Mg Magnesium 24.312											13 Al Aluminum 26.9815
19 K Potassium 39.102	20 Ca Calcium 40.08	21 Sc Scandium 44.956	22 Ti Titanium 47.90	23 V Vanadium 50.942	24 Cr Chromium 51.936	25 Mn Manganese 54.9380	26 Fe Iron 55.847	27 Co Cobalt 58.9332	28 Ni Nickel 58.71	29 Cu Copper 63.546	30 Zn Zinc 65.37	31 Ga Gallium 69.72
37 Rb Rubidium 85.47	38 Sr Strontium 87.62	39 Y Yttrium 88.905	40 Zr Zirconium 91.22	41 Nb Niobium 92.906	42 Mo Molybdenum 95.94	43 Tc Technetium (99)	44 Ru Ruthenium 101.07	45 Rh Rhodium 102.905	46 Pd Palladium 106.4	47 Ag Silver 107.868	48 Cd Cadmium 112.40	49 In Indium 114.82 / 50 Sn Tin 118.69
55 Cs Cesium 132.905	56 Ba Barium 137.34	57 La Lanthanum 138.91	72 Hf Hafnium 179.49	73 Ta Tantalum 180.948	74 W Tungsten 183.85	75 Re Rhenium 186.2	76 Os Osmium 190.2	77 Ir Iridium 192.2	78 Pt Platinum 195.09	79 Au Gold 196.967	80 Hg Mercury 200.59	81 Tl Thallium 204.37 / 82 Pb Lead 207.19 / 83 Bi Bismuth 208.980 / 84 Po Polonium (210)
87 Fr Francium (223)	88 Ra Radium (226)	89 Ac Actinium (227)	104 Rf Rutherfordium (261)	105 Db Dubnium (262)	106 Sg Seaborgium (266)	107 Bh Bohrium (264)	108 Hs Hassium (269)	109 Mt Meitnerium (268)	110 Uun Ununnilium (269)	111 Uuu Unununium (272)	112 Uub Ununbium (277)	

58 Ce Cerium 140.12	59 Pr Praseodymium 140.907	60 Nd Neodymium 144.24	61 Pm Promethium (145)	62 Sm Samarium 150.35	63 Eu Europium 151.96	64 Gd Gadolinium 157.25	65 Tb Terbium 158.924	66 Dy Dysprosium 162.50	67 Ho Holmium 164.930	68 Er Erbium 167.26	69 Tm Thulium 168.934	70 Yb Ytterbium 173.04	71 Lu Lutetium 174.97
90 Th Thorium 232.038	91 Pa Protactinium (231)	92 U Uranium 238.03	93 Np Neptunium (237)	94 Pu Plutonium (242)	95 Am Americium (243)	96 Cm Curium (247)	97 Bk Berkelium (247)	98 Cf Californium (251)	99 Es Einsteinium (254)	100 Fm Fermium (257)	101 Md Mendelevium (258)	102 No Nobelium (259)	103 Lr Lawrencium (260)

Figure 3-2: The metals.

Nonmetals

Except for the elements that border the stair-stepped line (more on those in a second), the elements to the right of the line, along with hydrogen, are classified as *nonmetals*. These elements are in Figure 3-3.

IA (1)		IVA (14)	VA (15)	VIA (16)	VIIA (17)	VIIIA (18)
						2 He Helium 4.0026
1 H Hydrogen 1.00797		6 C Carbon 12.01115	7 N Nitrogen 14.0067	8 O Oxygen 15.9994	9 F Flourine 18.9984	10 Ne Neon 20.183
			15 P Phosphorus 30.9738	16 S Sulfur 32.064	17 Cl Chlorine 35.453	18 Ar Argon 39.948
				34 Se Selenium 78.96	35 Br Bromine 79.904	36 Kr Krypton 83.80
					53 I Iodine 126.9044	54 Xe Xenon 131.30
						86 Rn Radon (222)

Figure 3-3: The nonmetals.

Nonmetals have properties opposite those of the metals. The nonmetals are brittle, aren't malleable or ductile, and are poor conductors of both heat and electricity. They tend to gain electrons in chemical reactions. Some nonmetals are liquids at room temperature.

Metalloids

The elements that border the stair-stepped line in the periodic table are classified as *metalloids,* and they're in Figure 3-4.

The metalloids, or *semimetals,* have properties that are somewhat of a cross between metals and nonmetals. They tend to be economically important because of their unique conductivity properties (they only partially conduct electricity), which make them valuable in the semiconductor and computer chip industry. (The term *Silicon Valley* doesn't refer to a valley covered in sand; silicon, one of the metalloids, is used in making computer chips.)

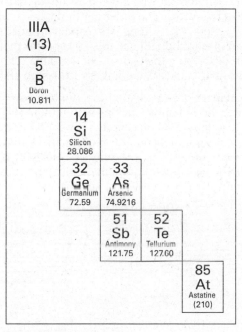

Figure 3-4: The metalloids.

Arranging elements by families and periods

The periodic table is composed of horizontal rows and vertical columns. Here's how they're named and numbered:

  ✔ **Periods:** The seven horizontal rows are called *periods.* The periods are numbered 1 through 7 on the left-hand side of the table (see Figure 3-1). Within each period, the atomic numbers increase from left to right.

Members of a period don't have very similar properties. Consider the first two members of period 3: sodium (Na) and magnesium (Mg). In reactions, they both tend to lose electrons (after all, they are metals), but sodium loses one electron, and magnesium loses two. Chlorine (Cl), down near the end of the period, tends to gain an electron (it's a nonmetal).

✔ **Families:** The vertical columns are called *groups,* or *families.* The families may be labeled at the top of the columns in one of two ways. The older method uses roman numerals and letters. Many chemists (especially academic ones like me) prefer and still use this method, so that's what I use in describing the features of the table. The newer method simply uses the numbers 1 through 18.

The members of a family do have similar properties. Consider the IA family, starting with lithium (Li) and going through francium (Fr) (don't worry about hydrogen, because it's unique, and it doesn't really fit anywhere). All these elements tend to lose only one electron in reactions. And all the members of the VIIA family tend to gain one electron.

Chapter 4

Nuclear Chemistry

● ●

In This Chapter

▶ Understanding radioactivity and radioactive decay

▶ Figuring out half-lives

▶ Knowing the basics of nuclear fission

▶ Taking a look at nuclear fusion

▶ Tracing the effects of radiation

● ●

*1*n one way or another, most of this book deals with chemical reactions. And when I talk about reactions, I'm really talking about how the valence electrons (the electrons in the outermost energy levels of atoms) are lost, gained, or shared. I mention very little about the nucleus of the atom because, to a very large degree, it's not involved in chemical reactions.

But in this chapter, I do discuss the nucleus and the changes it can undergo. I talk about radioactivity and the different ways an atom can decay. I discuss half-lives and show you how they're used in archeology. I also discuss nuclear fission and the hope that nuclear fusion holds for humankind. Finally, you get a quick glimpse of how radiation affects the cells in your body. Don't forget the lead shielding!

Seeing How the Atom's Put Together

To understand nuclear chemistry, you need to know the basics of atomic structure. Chapter 2 goes on and on about atomic structure, if you're interested. This section just provides a quickie brain dump.

The *nucleus,* that dense central core of the atom, contains both protons and neutrons. Electrons are outside the nucleus in energy levels. Protons have a positive charge, neutrons have no charge, and electrons have a negative charge. A neutral atom contains equal numbers of protons and electrons. But the number of neutrons within an atom of a particular element can vary. Atoms of the same element that have differing numbers of neutrons are called *isotopes.* Figure 4-1 shows the symbolization chemists use to represent a specific isotope of an element.

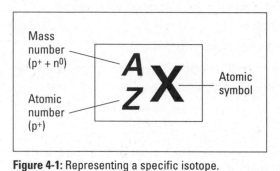

Figure 4-1: Representing a specific isotope.

In the figure, X represents the symbol of the element found on the periodic table, Z represents the *atomic number* (the number of protons in the nucleus), and A represents the *mass number* (the sum of the protons and neutrons in that particular isotope). If you subtract the atomic number from the mass number (A – Z), you get the number of neutrons in that particular isotope. A short way to show the same information is to simply use the element symbol (X) and the mass number (A) — for example, U-235.

Dealing with a Nuclear Breakup: Balancing Reactions

For purposes of this book, I define *radioactivity* as the spontaneous decay of an unstable nucleus. An unstable nucleus may break apart into two or more other particles with the release of some energy. This breaking apart can occur in a number of ways, depending on the particular atom that's decaying.

If you know what one of the particles of a radioactive decay will be, you can often predict the other particle. Doing so involves something called *balancing the nuclear reaction*. (A *nuclear reaction* is any reaction involving a change in nuclear structure.)

Balancing a nuclear reaction is really a fairly simple process. But before I explain it, here's how to represent a reaction:

Reactants → products

Reactants are the substances you start with, and products are the new substances being formed. The arrow, called a *reaction arrow*, indicates that a reaction has taken place.

For a nuclear reaction to be balanced, the sum of all the atomic numbers on the left-hand side of the reaction arrow must equal the sum of all the atomic numbers on the right-hand side of the arrow. The same is true for the sums of the mass numbers.

Here's an example: Suppose you're a scientist performing a nuclear reaction by bombarding a particular isotope of chlorine (Cl-35) with a neutron. You observe that an isotope of hydrogen, H-1, is created along with another isotope, and you want to figure out what the other isotope is. The equation for this example is

$$^{35}_{17}Cl + ^{1}_{0}n \rightarrow \underline{?} + ^{1}_{1}H$$

To figure out the unknown isotope (represented by *?*), you need to balance the equation. The sum of the atomic numbers on the left is 17 (17 + 0), so you want the sum of the atomic numbers on the right to equal 17, too. Right now, you have an atomic number of 1 on the right; 17 − 1 is 16, so that's the atomic number of the unknown isotope. This atomic number identifies the element as sulfur (S).

Now look at the mass numbers in the equation. The sum of the mass numbers on the left is 36 (35 + 1), and you want the sum of the mass numbers on the right to equal 36, too. Right now, you have a mass number of 1 on the right; 36 − 1 is 35, so that's the mass number of the unknown isotope. Now you know that the unknown isotope is a sulfur isotope (S-35). And here's what the balanced nuclear equation looks like:

$$^{35}_{17}Cl + ^{1}_{0}n \rightarrow ^{35}_{16}S + ^{1}_{1}H$$

This equation represents a *nuclear transmutation,* the conversion of one element into another. Nuclear transmutation is a process human beings control. S-35 is an isotope of sulfur that doesn't exist in nature. It's a human-made isotope. Alchemists, those ancient predecessors of chemists, dreamed of converting one element into another (usually lead into gold), but they were never able to master the process. Chemists are now able, sometimes, to convert one element into another.

Understanding Types of Natural Radioactive Decay

Certain isotopes are unstable: Their nuclei break apart, undergoing nuclear decay. Sometimes the product of that nuclear decay is unstable itself and undergoes nuclear decay, too. For example, when uranium-238 (U-238) initially decays, it produces thorium-234 (Th-234), which decays to protactinium-234 (Pa-234). The decay continues until, finally, after a total of 14 steps, lead-206 (Pb-206) is produced. Pb-206 is stable, and the decay sequence, or *series,* stops.

Before I show you how radioactive isotopes decay, first consider why a particular isotope decays. The nucleus has all those positively charged protons shoved together in an extremely small volume of space. All those protons are repelling each other. The forces that normally hold the nucleus together, the nuclear glue, sometimes can't do the job, and so the nucleus breaks apart, undergoing nuclear decay.

The neutron/proton ratio for a certain element must fall within a certain range for the element to be stable. All elements with 84 or more protons are unstable; they eventually undergo decay. Other isotopes with fewer protons in their nucleus are also radioactive. The radioactivity corresponds to the neutron/proton ratio in the atom. If the neutron/proton ratio is too high (there are too many neutrons or too few protons), the isotope is said to be *neutron rich* and is, therefore, unstable. Likewise, if the neutron/proton ratio is too low (there are too few neutrons or too many protons), the isotope is unstable. That's why some isotopes of an element are stable and others are radioactive.

Naturally occurring radioactive isotopes decay in three primary ways:

 ✔ Alpha particle emission

 ✔ Beta particle emission

 ✔ Gamma radiation emission

In addition, there are a couple of less common types of radioactive decay:

 ✔ Positron emission

 ✔ Electron capture

Alpha emission

An *alpha particle* is defined as a positively charged particle of a helium nucleus. This particle is composed of two protons and two neutrons, so it can be represented as a helium-4 atom. As an alpha particle breaks away from the nucleus of a radioactive atom, it has no electrons, so it has a +2 charge. Therefore, it's a positively charged particle of a helium nucleus. (Well, it's really a *cation*, a positively charged ion — see Chapter 2.)

But electrons are basically free — easy to lose and easy to gain. So normally, an alpha particle is shown with no charge because it very rapidly picks up two electrons and becomes a neutral helium atom instead of an ion.

Large, heavy elements, such as uranium and thorium, tend to undergo alpha emission. This decay mode relieves the nucleus of two units of positive charge (two protons) and four units of mass (two protons + two neutrons). What a process! Each time an alpha particle is emitted, four units of mass are lost.

Radon-222 (Rn-222) is another alpha particle emitter, as the following equation shows:

$$^{222}_{86}Rn \rightarrow\ ^{218}_{84}Po +\ ^{4}_{2}He$$

Here, radon-222 undergoes nuclear decay with the release of an alpha particle. The other remaining isotope must have a mass number of 218 (222 – 4) and an atomic number of 84 (86 – 2), which identifies the element as polonium (Po).

Beta emission

A *beta particle* is essentially an electron that's emitted from the nucleus. Iodine-131 (I-131), which is used in the detection and treatment of thyroid cancer, is a beta particle emitter:

$$^{131}_{53}I \rightarrow \, ^{131}_{54}Xe + \, ^{0}_{-1}e$$

Here, the iodine-131 gives off a beta particle (an electron), leaving an isotope with a mass number of 131 (131 – 0) and an atomic number of 54 (53 – (–1)). An atomic number of 54 identifies the element as xenon (Xe). Notice that the mass number doesn't change in going from I-131 to Xe-131, but the atomic number increases by one.

In *beta emission,* a neutron in the nucleus is converted (decayed) into a proton and an electron, and the electron is emitted from the nucleus as a beta particle. Isotopes with a high neutron/proton ratio often undergo beta emission, because this decay mode allows the number of neutrons to be decreased by one and the number of protons to be increased by one, thus lowering the neutron/proton ratio.

Gamma emission

Alpha and beta particles have the characteristics of matter: They have definite masses, occupy space, and so on. However, because no mass change is associated with gamma emission, I refer to gamma emission as *gamma radiation emission*.

Gamma radiation is similar to X-rays — high energy, short wavelength radiation. Gamma radiation commonly accompanies both alpha and beta emission, but it's usually not shown in a balanced nuclear reaction. Some isotopes, such as cobalt-60 (Co-60), give off large amounts of gamma radiation. Co-60 is used in the radiation treatment of cancer. The medical personnel focus gamma rays on the tumor, thus destroying it.

Positron emission

Although positron emission doesn't occur with naturally occurring radioactive isotopes, it does occur naturally in a few human-made ones. A *positron* is essentially an electron that has a positive charge instead of a negative charge.

A positron is formed when a proton in the nucleus decays into a neutron and a positively charged electron. The positron is then emitted from the nucleus. This process occurs in a few isotopes, such as potassium-40 (K-40), as the following equation shows:

$$_{19}^{40}K \rightarrow _{18}^{40}Ar + _{+1}^{0}e$$

The K-40 emits the positron, leaving an element with a mass number of 40 (40 − 0) and an atomic number of 18 (19 − 1). An isotope of argon (Ar), Ar-40, has been formed.

Electron capture

Electron capture is a rare type of nuclear decay in which the nucleus captures an electron from the innermost energy level (the 1s — see Chapter 3). This electron combines with a proton to form a neutron. The atomic number decreases by one, but the mass number stays the same. The following equation shows the electron capture of polonium-204 (Po-204):

$$_{84}^{204}Po + _{-1}^{0}e \rightarrow _{83}^{204}Bi + x - rays$$

The electron combines with a proton in the polonium nucleus, creating an isotope of bismuth (Bi-204).

The capture of the 1s electron leaves a vacancy in the 1s orbitals. Electrons drop down to fill the vacancy, releasing energy not in the visible part of the electromagnetic spectrum but in the X-ray portion.

Half-Lives and Radioactive Dating

If you could watch a single atom of a radioactive isotope, U-238, for example, you wouldn't be able to predict when that particular atom might decay. It may take a millisecond, or it may take a century. There's simply no way to tell.

But if you have a large enough sample — what mathematicians call a *statistically significant sample size* — a pattern begins to emerge. It takes a certain amount of time for half the atoms in a sample to decay. It then takes the same amount of time for half the remaining radioactive atoms to decay, and the same amount of time for half of those remaining radioactive atoms to decay, and so on. The amount of time it takes for one-half of a sample to decay is called the *half-life* of the isotope, and it's given the symbol $t_{1/2}$. Table 4-1 shows this process.

Table 4-1	Half-Life Decay of a Radioactive Isotope
Number of Half-Lives	*Percent of the Radioactive Isotope Remaining*
0	100.00
1	50.00
2	25.00
3	12.50
4	6.25
5	3.13
6	1.56
7	0.78
8	0.39
9	0.20
10	0.10

Calculating remaining radioactivity

The half-life decay of radioactive isotopes is not linear. For example, you can't find the remaining amount of an isotope at 7.5 half-lives by finding the midpoint between 7 and 8 half-lives.

If you want to find times or amounts that are not associated with a simple multiple of a half-life, you can use this equation:

$$\ln\left(\frac{(N_0)}{N}\right) = \left(\frac{0.6963}{t_{1/2}}\right)t$$

In the equation, *ln* stands for the *natural logarithm* (the base *e* log, not the base 10 log; it's that *ln* button on your calculator, not the *log* button). *No* is the amount of radioactive isotope that you start with (in grams, as a percentage, in the number of atoms, and so on), *N* is the amount of radioisotope left at some time *(t)*, and $t_{1/2}$ is the half-life of the radioisotope. If you know the half life and the amount of the radioactive isotope that you start with, you can use this equation to calculate the amount remaining radioactive at any time.

Radioactive dating

A useful application of half-lives is *radioactive dating*. Carbon-14 (C-14), a radioactive isotope of carbon, is produced in the upper atmosphere by cosmic radiation. The primary carbon-containing compound in the atmosphere is carbon dioxide, and a very small amount of carbon dioxide contains C-14. Plants absorb C-14 during photosynthesis, so C-14 is incorporated into the cellular structure of plants. Plants are then eaten by animals, making C-14 a part of the cellular structure of all living things.

As long as an organism is alive, the amount of C-14 in its cellular structure remains constant. But when the organism dies, the amount of C-14 begins to decrease. Scientists know the half-life of C-14 (5,730 years), so they can figure out how long ago the organism died.

Radioactive dating using C-14 has been used to determine the age of skeletons found at archeological sites. It was also used to date the *Shroud of Turin,* a piece of linen in the shape of a burial cloth that contains an image of a man. Many thought that it was the burial cloth of Jesus, but in 1988, radiocarbon dating determined that the cloth dated from around 1200–1300 CE.

Carbon-14 dating can only determine the age of something that was once alive. It can't determine the age of a moon rock or a meteorite. For nonliving substances, scientists use other isotopes, such as potassium-40.

Breaking Elements Apart with Nuclear Fission

In the 1930s, scientists discovered that some nuclear reactions can be initiated and controlled. Scientists usually accomplished this task by bombarding a large isotope with a second, smaller one — commonly a neutron. The collision caused the larger isotope to break apart into two or more elements, which is called *nuclear fission.* The following equation shows the nuclear fission of uranium-235:

$$^{235}_{92}U + {}^1_0n \rightarrow {}^{142}_{56}Ba + {}^{91}_{36}Kr + 3{}^1_0n$$

Mass defect: Where does all that energy come from?

Nuclear fission reactions release a lot of energy. Where does the energy come from? Well, if you make *very* accurate measurements of the masses of all the atoms and subatomic particles you start with and all the atoms and subatomic particles you end up with, you find that some mass is "missing." Matter disappears during the nuclear reaction. This loss of matter is called the *mass defect.* The missing matter is converted into energy.

You can actually calculate the amount of energy produced during a nuclear reaction with a fairly simple equation developed by Einstein: $E = mc^2$. In this equation, E is the amount of energy produced, m is the "missing" mass, or mass defect, and c is the speed of light in a vacuum, which is a rather large number. The speed of light is squared, making that part of the equation a very large number that, even when multiplied by a small amount of mass, yields a large amount of energy.

Chain reactions and critical mass

Take a look at the equation for the fission of uranium-235 (U-235):

$$^{235}_{92}U + ^{1}_{0}n \rightarrow ^{142}_{56}Ba + ^{91}_{36}Kr + 3^{1}_{0}n$$

Notice that one neutron is used, but three are produced. These three neutrons, if they encounter other U-235 atoms, can initiate other fissions, producing even more neutrons. It's the old domino effect — or in terms of nuclear chemistry, it's a continuing cascade of nuclear fissions called a *chain reaction*. Figure 4-2 shows the chain reaction of U-235.

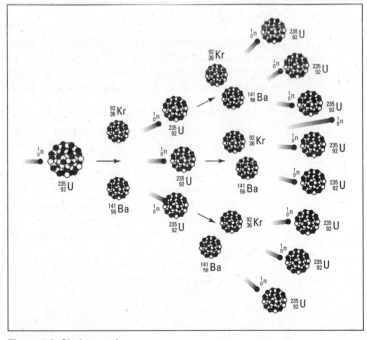

Figure 4-2: Chain reaction.

A chain reaction depends on the release of more neutrons than are used during the nuclear reaction. If you were to write the equation for the nuclear fission of U-238, the more abundant isotope of uranium, you'd use one neutron and get only one back out. So you can't have a chain reaction with U-238. But isotopes that produce an excess of neutrons in their fission support a chain reaction. This type of isotope is said to be *fissionable,* and only two main fissionable isotopes are used during nuclear reactions: U-235 and plutonium-239 (Pu-239).

Critical mass is the minimum amount of fissionable matter you need to support a self-sustaining chain reaction. That amount is related to those neutrons. If the sample is small, then the neutrons are likely to shoot out of the sample before hitting a U-235 nucleus. If they don't hit a U-235 nucleus, no extra electrons and no energy are released. The reaction just fizzles. Anything less than the critical mass is called *subcritical.*

Coming Together with Nuclear Fusion

Soon after researchers discovered the fission process, they discovered another process, called fusion. Fusion is essentially the opposite of fission. In fission, a heavy nucleus is split into smaller nuclei. With *fusion,* lighter nuclei are fused into a heavier nucleus.

The fusion process is the reaction that powers the sun. In a series of nuclear reactions in the sun, four hydrogen-1 (H-1) isotopes are fused into a helium-4 (He-4) with the release of a tremendous amount of energy. Here on Earth, people use two other isotopes of hydrogen: H-2, called *deuterium,* and H-3, called *tritium.* (Deuterium is a minor isotope of hydrogen, but it's still relatively abundant. Tritium doesn't occur naturally, but people can easily produce it by bombarding deuterium with a neutron.)

The following equation shows the fusion reaction:

$$^3_1H + {}^2_1H \rightarrow {}^4_2He + {}^1_0n$$

Chapter 5

Ionic Bonding

● ●

In This Chapter

▶ Finding out why and how ions are formed

▶ Understanding how ions create chemical bonds

▶ Deciphering the formulas of ionic compounds

▶ Naming ionic compounds

▶ Connecting conductivity and ionic bonds

● ●

*I*n this chapter, I introduce you to ionic bonding, the type of bonding that holds salts together. I discuss simple ions and polyatomic ions: how they form and how they combine. I also show you how to predict the formulas of ionic compounds and how chemists detect ionic bonds.

Forming Ions: Making Satisfying Electron Trades

In nature, achieving a filled (complete) valence energy level is a driving force of chemical reactions, because when that energy level is full, elements become stable, or "satisfied" — stable elements don't lose, gain, or share electrons.

The *noble gases* — the VIIIA elements on the periodic table — are extremely nonreactive because their valence energy level (outermost energy level) is filled. However, the other elements in the A families on the periodic table do gain, lose, or share valence electrons to fill their valence energy level and become satisfied.

Because filling the valence energy level usually involves filling the outermost s and p orbitals, it's sometimes called the *octet rule* — elements gain, lose, or share electrons to reach a full octet (eight valence electrons: two in the s orbital and six in the p orbital).

In this section, I explain how atoms gain or lose electrons to form ions and achieve stability. I also explain how ions can consist of single atoms or a group of atoms. (For info on achieving stability by sharing electrons, flip to Chapter 6.)

Gaining and losing electrons

When an atom gains or loses an electron, it develops a charge and becomes an *ion*. In general, the loss or gain of one, two, or sometimes even three electrons can occur, but an element doesn't lose or gain more than three electrons.

Losing an electron to become a cation: Sodium

Ions that have a positive charge due to the loss of electrons are called *cations*. In general, a cation is smaller than its corresponding atom. Why? The filled energy level determines the size of an atom or ion, and a cation gives up enough electrons to lose an entire energy level.

Consider sodium, an alkali metal and a member of the IA family on the periodic table. Sodium has 1 valence electron and 11 total electrons, because its atomic number is 11. It has an electron configuration of $1s^2 2s^2 2p^6 3s^1$. (See Chapter 2 for a review of electron configurations.)

By the octet rule, sodium becomes stable when it has eight valence electrons. Two possibilities exist for sodium to become stable: It can gain seven more electrons to fill energy level 3, or it can lose the one 3s electron so that energy level 2 (which is already filled at eight electrons) becomes the valence energy level.

So to gain stability, sodium loses its 3s electron. At this point, it has 11 protons (11 positive charges) and 10 electrons (10 negative charges). The once-neutral sodium atom now has a single positive charge [11 (+) plus 10 (−) equals 1+]. It's now an *ion,* an atom that has a charge due to the loss or gain of electrons. You can write an electron configuration for the sodium cation:

Na⁺: $1s^2 2s^2 2p^6$

Note that if an ion simply has 1 unit of charge, positive or negative, you normally don't write the 1; you just use the plus or minus symbol, with the 1 being understood.

Atoms that have matching electron configurations are *isoelectronic* with each other. The positively charged sodium ion (cation) has the same electron configuration as neon, so it's isoelectronic with neon. So does sodium become neon by losing an electron? No. Sodium still has 11 protons, and the number of protons determines the identity of the element.

There's a difference between the neutral sodium atom and the sodium cation: one electron. As a result, their chemical reactivities are different and their sizes are different. Because sodium loses an entire energy level to change from a neutral atom to a cation, the cation is smaller.

Gaining an electron to become an anion: Chlorine

Ions with a negative charge due to the gain of electrons are called *anions*. In general, an anion is slightly larger than its corresponding atom because the protons have to attract one or more extra electrons. The attractive force is slightly reduced, so the electrons are free to move outward a little.

Chlorine, a member of the halogen family — the VIIA family on the periodic table — often forms anions. It has seven valence electrons and a total of 17 electrons, and its electron configuration is $1s^2 2s^2 2p^6 3s^2 3p^5$. So to obtain its full octet, chlorine must lose the seven electrons in energy level 3 or gain one at that level.

Because elements don't gain or lose more than three electrons, chlorine must gain a single electron to fill energy level 3. At this point, chlorine has 17 protons (17 positive charges) and 18 electrons (18 negative charges). So chlorine becomes an ion with a single negative charge (Cl⁻). The neutral chlorine atom becomes the *chloride ion*. The electronic configuration for the chloride anion is

Cl⁻: $1s^2 2s^2 2p^6 3s^2 3p^6$

The chloride anion is isoelectronic with argon. The chloride anion is also slightly larger than the neutral chlorine atom. To complete the octet, the one electron gained went into energy level 3. But now there are 17 protons attracting 18 electrons, so the electrons can move outward a bit.

Looking at charges on single-atom ions

In the periodic table, the roman numerals at the top of the A families show the number of valence electrons in each element. Because atoms form ions to achieve full valence energy levels, that means you can often use an element's position in the periodic table to figure out what kind of charge an ion normally has. Here's how to match up the A families with the ions they form:

✔ **IA family (alkali metals):** Each element has one valence electron, so it loses a single electron to form a cation with a 1+ charge.

✔ **IIA family (alkaline earth metals):** Each element has two valence electrons, so it loses two electrons to form a 2+ cation.

✔ **IIIA family:** Each element has three valence electrons, so it loses three electrons to form a 3+ cation.

✔ **VA family:** Each element has five valence electrons, so it gains three electrons to form an anion with a 3– charge.

✔ **VIA family:** Each element has six valence electrons, so it gains two electrons to form an anion with a 2– charge.

✔ **VIIA family (halogens):** Each element has seven valence electrons, so it gains a single electron to form an anion with a 1– charge.

Determining the number of electrons that members of the transition metals (the B families) lose is more difficult. In fact, many of these elements lose a varying number of electrons so that they form two or more cations with different charges.

Seeing some common one-atom ions

Table 5-1 shows the family, element, ion name, and ion symbol for some common monoatomic (one-atom) cations.

Table 5-1   Common Monoatomic Cations

Family	Element	Ion Name	Ion Symbol
IA	Lithium	Lithium cation	Li^+
	Sodium	Sodium cation	Na^+
	Potassium	Potassium cation	K^+
IIA	Beryllium	Beryllium cation	Be^{2+}
	Magnesium	Magnesium cation	Mg^{2+}
	Calcium	Calcium cation	Ca^{2+}
	Strontium	Strontium cation	Sr^{2+}
	Barium	Barium cation	Ba^{2+}
IB	Silver	Silver cation	Ag^+
IIB	Zinc	Zinc cation	Zn^{2+}
IIIA	Aluminum	Aluminum cation	Al^{3+}

Table 5-2 gives the same information for some common monoatomic anions.

Table 5-2   Common Monoatomic Anions

Family	Element	Ion Name	Ion Symbol
VA	Nitrogen	Nitride anion	N^{3-}
	Phosphorus	Phosphide anion	P^{3-}
VIA	Oxygen	Oxide anion	O^{2-}
	Sulfur	Sulfide anion	S^{2-}
VIIA	Fluorine	Fluoride anion	F^-
	Chlorine	Chloride anion	Cl^-
	Bromine	Bromide anion	Br^-
	Iodine	Iodide anion	I^-

Possible charges: Naming ions with multiple oxidation states

The electrical charge that an atom achieves is sometimes called its *oxidation state*. Many of the transition metal ions (the B families) have varying oxidation states because these elements can vary in how many electrons they lose. Table 5-3 shows some common transition metals that have more than one oxidation state.

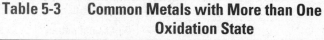

Table 5-3   Common Metals with More than One Oxidation State

Family	Element	Ion Name	Ion Symbol
VIB	Chromium	Chromium (II) or chromous	Cr^{2+}
		Chromium (III) or chromic	Cr^{3+}
VIIB	Manganese	Manganese (II) or manganous	Mn^{2+}
		Manganese (III) or manganic	Mn^{3+}
VIIIB	Iron	Iron (II) or ferrous	Fe^{2+}
		Iron (III) or ferric	Fe^{3+}
	Cobalt	Cobalt (II) or cobaltous	Co^{2+}
		Cobalt (III) or cobaltic	Co^{3+}
IB	Copper	Copper (I) or cuprous	Cu^{+}
		Copper (II) or cupric	Cu^{2+}
IIB	Mercury	Mercury (I) or mercurous	Hg_2^{2+}
		Mercury (II) or mercuric	Hg^{2+}
IVA	Tin	Tin (II) or stannous	Sn^{2+}
		Tin (IV) or stannic	Sn^{4+}
	Lead	Lead (II) or plumbous	Pb^{2+}
		Lead (IV) or plumbic	Pb^{4+}

Notice that these cations can have more than one name. Here are two ways to name cations of elements that have more than one oxidation state:

 ✔ **Current method:** Use the metal name, such as chromium, followed by the ionic charge written as a roman numeral in parentheses, such as (II). For example, Cr^{2+} is *chromium (II)* and Cr^{3+} is *chromium (III)*.

 ✔ **Traditional method:** An older way of naming ions uses *-ous* and *-ic* endings. When an element has more than one ion, do the following:

- Give the ion with the lower oxidation state (lower numerical charge, ignoring the + or –) an *-ous* ending.

- Give the ion with the higher oxidation state (higher numerical charge) an *-ic* ending.

So for chromium, the Cr^{2+} ion is named *chromous* and the Cr^{3+} ion is named *chromic*.

Grouping atoms to form polyatomic ions

Ions can be *polyatomic,* composed of a group of atoms. For example, take a look at Table 5-3 in the preceding section. Notice anything about the mercury (I) ion? Its ion symbol, Hg_2^{2+}, shows that two mercury atoms are bonded together. This group has a 2+ charge, with each mercury cation having a 1+ charge. The mercurous ion is classified as a polyatomic ion.

Similarly, the symbol for the sulfate ion, SO_4^{2-}, indicates that one sulfur atom and four oxygen atoms are bonded together and that the whole polyatomic ion has two extra electrons: a 2– charge.

Polyatomic ions are treated the same as monoatomic ions (see "Naming ionic compounds," later in this chapter). Table 5-4 lists some important polyatomic ions.

Table 5-4	Some Important Polyatomic Ions
Ion Name	*Ion Symbol*
Sulfite	SO_3^{2-}
Sulfate	SO_4^{2-}
Thiosulfate	$S_2O_3^{2-}$
Bisulfate (or hydrogen sulfate)	HSO_4^-
Nitrite	NO_2^-
Nitrate	NO_3^-
Hypochlorite	ClO^-
Chlorite	ClO_2^-

(continued)

Table 5-4 *(continued)*

Ion Name	Ion Symbol
Chlorate	ClO_3^-
Perchlorate	ClO_4^-
Chromate	CrO_4^{2-}
Dichromate	$Cr_2O_7^{2-}$
Arsenite	AsO_3^{3-}
Arsenate	AsO_4^{3-}
Phosphate	PO_4^{3-}
Hydrogen phosphate	HPO_4^{2-}
Dihydrogen phosphate	$H_2PO_4^-$
Carbonate	CO_3^{2-}
Bicarbonate (or hydrogen carbonate)	HCO_3^-
Cyanide	CN^-
Cyanate	OCN^-
Thiocyanate	SCN^-
Peroxide	O_2^{2-}
Hydroxide	OH^-
Acetate	$C_2H_3O_2^-$
Oxalate	$C_2O_4^{2-}$
Permanganate	MnO_4^-
Ammonium	NH_4^+
Mercury (I)	Hg_2^{2+}

Creating Ionic Compounds

Ionic bonding, the bonding that holds the cations and anions together, is one of the two major types of bonding in chemistry. (I describe the other type, *covalent bonding,* in Chapter 6.)

An ionic bond occurs between a metal and a nonmetal. The metal loses electrons (to becomes a positively charged cation), and a nonmetal gains those electrons (to become a negatively charged anion). The ions have opposite charges, so they're attracted to each other. This attraction draws them together into a compound.

In this section, you look at how ionic bonding works, and you see how to write formulas for and name ionic compounds.

Making the bond: Sodium metal + chlorine gas = sodium chloride

The transfer of an electron creates ions — cations (positive charge) and anions (negative charge). Opposite charges attract each other, so the cations and anions may come together through an ionic bond. An *ionic bond* is a chemical bond (a strong attractive force that keeps two chemical elements together) that comes from the *electrostatic attraction* (attraction of opposite charges) between cations and anions. Together, the ions form a compound.

For instance, sodium, a metal, can fill its octet and achieve stability by losing an electron. Chlorine, a nonmetal, can fill its octet by gaining an electron. (See the earlier section "Gaining and losing electrons" for details on the octet rule.) If the two are in the same container, then the electron that sodium loses can be the same electron that chlorine gains. The Na^+ cation attracts the Cl^- anion and forms the compound NaCl, sodium chloride.

Compounds that have ionic bonds are commonly called *salts*. In sodium chloride — table salt — a crystal is formed in which each sodium cation is surrounded by six different chloride anions and each chloride anion is surrounded by six different sodium cations.

Different types of salts have different crystal structures. Cations and anions can have more than one unit of positive or negative charge if they lose or gain more than one electron. In this fashion, many different kinds of salts are possible.

Figuring out the formulas of ionic compounds

When an ionic compound is formed, the cation and anion attract each other, resulting in a salt. This section shows you how to write the formula of that salt.

Balancing charges: Magnesium and bromine

Suppose you want to know the *formula*, or composition, of a compound that results from reacting a metal and a nonmetal. You start by putting the two atoms side by side, with the metal on the left. Then you add their charges.

Figure 5-1 shows this process for magnesium and bromine. (Forget about the crisscrossing lines for now. I explain them in the upcoming section "Using the crisscross rule.")

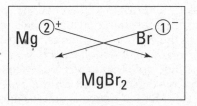

Figure 5-1: Figuring the formula of magnesium bromide.

The electron configurations for magnesium and bromine are

Magnesium (Mg): $1s^2 2s^2 2p^6 3s^2$

Bromine (Br): $1s^2 2s^2 2p^6 3s^2 3p^6 4s^2 3d^{10} 4p^5$

Magnesium, an alkaline earth metal, has two valence electrons that it loses to form a cation with a 2+ charge. The electron configuration for the magnesium cation is

Mg^{2+}: $1s^2 2s^2 2p^6$

Bromine, a halogen, has seven valence electrons, so it gains one electron to complete its octet (eight valence electrons) and form the bromide anion with a 1– charge. The electron configuration for the bromide anion is

Br^{1-}: $1s^2 2s^2 2p^6 3s^2 3p^6 4s^2 3d^{10} 4p^6$

When writing the formula of a compound, the compound must be neutral. That is, it needs to have equal numbers of positive and negative charges. So after writing the atoms, you need to balance their charges.

The magnesium ion has a 2+, so it requires two bromide anions, each with a single negative charge, to balance the two positive charges of magnesium. So the formula of the compound that results from reacting magnesium with bromine is $MgBr_2$.

Using the crisscross rule

A quick way to determine the formula of an ionic compound is to use the *crisscross rule:* Take the numerical value of the metal ion's superscript (forget about the charge symbol) and move it to the bottom right-hand side of the nonmetal's symbol — as a subscript. Then take the numerical value of the nonmetal's superscript and make it the subscript of the metal. (Note that if the numerical value is 1, it's just understood and not shown.)

To see how to use this rule, look back at Figure 5-1. For magnesium and bromine, you make magnesium's 2 a subscript of bromine and make bromine's 1 a subscript of magnesium (but because it's 1, you don't show it). You get the formula $MgBr_2$.

So what happens if you react aluminum and oxygen? Figure 5-2 shows the crisscross rule for this reaction. You get Al_2O_3.

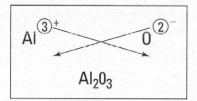

Figure 5-2: Figuring out the formula of aluminum oxide.

Compounds involving polyatomic ions work in exactly the same way. For example, here's the compound made from the ammonium cation (NH_4^+) and the sulfide anion (S^{2-}):

$(NH_4)_2S$

Notice that because you need two ammonium ions (two positive charges) to neutralize the two negative charges of the sulfide ion, you enclose the ammonium ion in parentheses and add a subscript 2.

After you use the crisscross rule, reduce all the subscripts by a common factor, if possible, to get the right formula.

For example, suppose that you want to write the compound formed when calcium reacts with oxygen. Calcium, an alkaline earth metal, forms a 2+ cation, and oxygen forms a 2– anion. So you may predict that the formula is

Ca_2O_2

But you need to divide each subscript by 2 to get the correct formula:

CaO

Naming ionic compounds

When you name inorganic compounds, you write the name of the metal first and then the nonmetal, adding an -*ide* ending to the nonmetal (for compounds involving monatomic ions).

Suppose, for example, that you want to name Li_2S, the compound that results from the reaction of lithium and sulfur. You first write the name of the metal, lithium, and then write the name of the nonmetal, adding an -*ide* ending so that *sulfur* becomes *sulfide*:

Li_2S: Lithium sulfide

Ionic compounds involving polyatomic ions follow the same basic rule: Write the name of the metal first, and then simply add the name of the nonmetal. However, with polyatomic anions, it's not necessary to add the -ide ending:

$(NH_4)_2CO_3$: Ammonium carbonate

K_3PO_4: Potassium phosphate

Dealing with multiple oxidation states

When the metal involved is a transition metal with more than one oxidation state (see Table 5-3, earlier in the chapter), there can be more than one way to correctly name the compound, based on how you name the metal.

For example, suppose that you want to name the compound formed between the Fe^{3+} cation and the cyanide ion, CN^-. The preferred method is to use the metal name followed in parentheses by the ionic charge written as a roman numeral: iron (III). But an older naming method, which is still sometimes used (so it's a good idea to know it), is to use -*ous* and -*ic* endings.

The ion with the lower oxidation state (lower numerical charge, ignoring the + or –) gets an -*ous* ending, and the ion with the higher oxidation state (higher numerical charge) gets an -*ic* ending. So because Fe^{3+} has a higher oxidation state than Fe^{2+}, it's called a *ferric ion*.

After you write the name of the metal, name the nonmetal. So the compound $Fe(CN)_3$ can be named

$Fe(CN)_3$: iron(III) cyanide, or ferric cyanide

Getting names from formulas and formulas from names

Sometimes figuring out the charge on an ion can be a little challenging (and fun), so take a look at how to name $FeNH_4(SO_4)_2$. I show you earlier in Table 5-4 that the sulfate ion has a 2– charge, and from the formula you can see that there are two of these ions. Therefore, you have a total of four negative charges. Table 5-4 also indicates that the ammonium ion has a 1+ charge, so you can figure out the charge on the iron cation:

Ion	*Charge*
Fe	?
NH_4	1+
$(SO_4)_2$	$(2-) \times 2$

Because you have a 4– charge for the sulfates and a 1+ for the ammonium, the iron must be a 3+ to make the compound neutral. So the iron is in the iron (III), or *ferric,* oxidation state. You can name the compound:

$FeNH_4(SO4)_2$: Iron (III) ammonium sulfate, or ferric ammonium sulfate

And finally, if you have the name, you can derive the formula and the charge on the ions. For example, suppose that you're given the name *cuprous oxide*. You know that the cuprous ion is Cu^+ and the oxide ion is O^{2-}. Applying the crisscross rule (from the earlier section "Using the crisscross rule"), you get the following formula:

Cuprous oxide: Cu_2O

Bonding Clues: Electrolytes and Nonelectrolytes

Scientists can get some good clues about the type of bonding in a compound by discovering whether a substance is an electrolyte or a nonelectrolyte. Ionically bonded substances act as electrolytes, but covalently bonded compounds, in which no ions are present, are commonly nonelectrolytes.

Electrolytes are substances that conduct electricity in the molten state or when dissolved in water. For instance, sodium chloride is an electrolyte because it conducts an electrical current when dissolved in water. If you were to melt pure NaCl (which requires a lot of heat!) and then check the conductivity of the molten salt, you'd find that the molten table salt also conducts electricity. In the molten state, the NaCl ions are free to move and carry electrons, just as they are in the saltwater solution.

Substances that don't conduct electricity when in these states are called *nonelectrolytes*. Table sugar, or sucrose, is a good example of a nonelectrolyte. You can dissolve sugar in water or melt it, but it won't have conductivity. No ions are present to transfer the electrons.

Chapter 6

Covalent Bonding

● ●

In This Chapter

▶ Seeing how one hydrogen atom bonds to another hydrogen atom

▶ Defining covalent bonding

▶ Finding out about the different types of chemical formulas

▶ Taking a look at polar covalent bonding and electronegativity

● ●

*W*hat holds together sugar, vinegar, and even DNA? Not ionic bonds! In this chapter, I discuss the other major type of bonding: covalent bonding. I explain the basics with an extremely simple covalent compound, hydrogen.

Covalent Bond Basics

Atoms form compounds to achieve a filled valence energy level (see Chapter 2 for more on energy levels). But instead of achieving it by gaining or losing electrons, as in ionic bonding (Chapter 5), the atoms in some compounds share electrons. That's the basis of a *covalent bond*.

Sharing electrons: A hydrogen example

Hydrogen is number 1 on the periodic table — upper-left corner. Hydrogen has one valence electron. It'd love to gain another electron to fill its 1s energy level, which would make it *isoelectronic* with helium (because the two would have the same electronic configuration), the nearest noble gas. Energy level 1 can hold only two electrons in the 1s orbital, so gaining another electron would fill it. That's the driving force of hydrogen: filling the valence energy level and achieving the same electron arrangement as the nearest noble gas.

Why atoms have to share

Why can't the simple gain or loss of electrons explain the stability of H_2? Imagine one hydrogen atom transferring its single electron to another hydrogen atom. The hydrogen atom receiving the electron fills its valence shell and reaches stability while becoming an anion (H^-). However, the other hydrogen atom now has no electrons (H^+) and moves further away from stability. This process of electron loss and gain simply won't happen, because the goal of both atoms is to fill their valence energy levels. So the H_2 compound can't result from the loss or gain of electrons.

What *can* happen is that the two atoms can share their electrons. At the atomic level, this sharing is represented by the electron orbitals (sometimes called *electron clouds*) overlapping. The two electrons (one from each hydrogen atom) "belong" to both atoms. Each hydrogen atom feels the effect of the two electrons; each has, in a way, filled its valence energy level. A *covalent bond* is formed — a chemical bond that comes from the sharing of one or more electron pairs between two atoms.

That's why the hydrogen found in nature is often not comprised of an individual atom. It's primarily found as H_2, a *diatomic* (two-atom) compound. Taken one step further, because a *molecule* is a combination of two or more atoms, H_2 is called a *diatomic molecule*.

In addition to hydrogen, six other elements are found in nature in the diatomic form: oxygen (O_2), nitrogen (N_2), fluorine (F_2), chlorine (Cl_2), bromine (Br_2), and iodine (I_2). So when I talk about oxygen gas or liquid bromine, I'm talking about the diatomic compound (diatomic molecule).

Representing covalent bonds

The overlapping of the electron orbitals and the sharing of an electron pair is represented in Figure 6-1a.

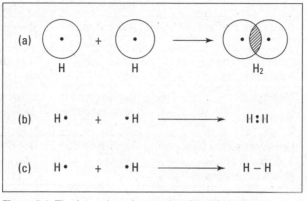

Figure 6-1: The formation of a covalent bond in hydrogen.

Another way to represent this process is through the use of an *electron-dot formula*. In this type of formula, valence electrons are represented as dots surrounding the atomic symbol, and the shared electrons are shown between the two atoms involved in the covalent bond. Figure 6-1b shows the electron-dot formula representations of H_2.

Most of the time, I use a slight modification of the electron-dot formula called the *Lewis structural formula;* it's basically the same as the electron-dot formula, but the shared pair of electrons (the covalent bond) is represented by a dash. Figure 6-1c shows the Lewis structural formula.

Comparing covalent bonds with other bonds

The properties of ionic and covalent compounds are different. Table 6-1 shows how the compounds compare. (*Note:* For the classification between metals and nonmetals, see Chapter 3.)

Table 6-1	Properties of Ionic and Covalent Compounds	
Property	Ionic Compounds (Salts)	Covalent Compounds
Bonds occur between	A metal and a nonmetal	Two nonmetals
State of the compound at room temperature	Usually solid	Can be solid, liquid, or gas
Melting point	Higher than for covalent compounds	Lower than for ionic compounds
Electrolytes (they form ions and conduct electricity when dissolved) or nonelectrolytes	Tend to be electrolytes	Tend to be nonelectrolytes

I know just what you're thinking: If metals react with non-metals to form ionic bonds, and nonmetals react with other nonmetals to form covalent bonds, do metals react with other metals? The answer is yes and no.

Metals don't really react with other metals to form compounds. Instead, the metals combine to form *alloys,* solutions of one metal in another. But there is such a situation as metallic bonding, and it's present in both alloys and pure metals. In *metallic bonding,* the valence electrons of each metal atom are donated to an electron pool, commonly called a *sea of electrons,* and are shared by all the atoms in the metal. These valence electrons are free to move throughout the sample instead of being tightly bound to an individual metal nucleus. The ability of the valence electrons to flow throughout the entire metal sample is why metals tend to be conductors of electricity and heat.

Dealing with multiple bonds

I define covalent bonding as the sharing of one *or more* electron pairs. In hydrogen and most other diatomic molecules, only one electron pair is shared. But in many covalent bonding situations, the atoms share more than one electron pair.

For instance, nitrogen (N_2) is a diatomic molecule in which the atoms share more than one electron pair.

The nitrogen atom is in the VA family on the periodic table, meaning that it has five valence electrons (see Chapter 3 for details on the periodic table). So nitrogen needs three more valence electrons to complete its octet. A nitrogen atom can fill its octet by sharing three electrons with another nitrogen atom, forming three covalent bonds, a so-called *triple bond*. Figure 6-2 shows the triple bond formation of nitrogen.

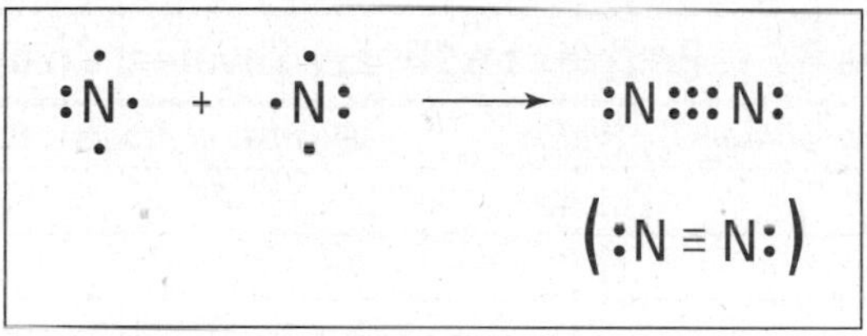

$$:\!N\cdot \ + \ \cdot N\!: \ \longrightarrow \ :\!N\!:\!:\!:\!N\!:$$

$$\left(:\!N \equiv N\!:\right)$$

Figure 6-2: Triple bond formation in N_2.

A triple bond isn't quite three times as strong as a single bond, but it's a very strong bond. In fact, the triple bond in nitrogen is one of the strongest bonds known. This strong bond is what makes nitrogen very stable and resistant to reaction with other chemicals. It's also why many explosive compounds (such as TNT and ammonium nitrate) contain nitrogen: When these compounds break apart in a chemical reaction, nitrogen gas (N_2) is formed, and a large amount of energy is released.

Carbon dioxide (CO_2) is another example of a compound containing a multiple bond. Carbon can react with oxygen to form carbon dioxide. Carbon has four valence electrons, and oxygen has six. Carbon can share two of its valence electrons with each of the two oxygen atoms, forming two double bonds. Figure 6-3 shows these double bonds.

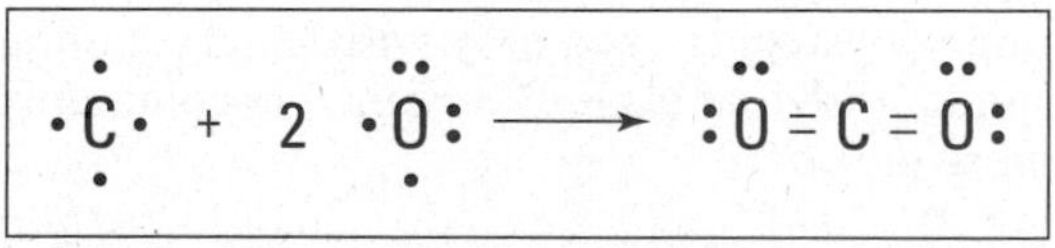

$$\cdot C\cdot \ + \ 2 \ \cdot \overset{\cdot\cdot}{O}\!: \ \longrightarrow \ :\!\overset{\cdot\cdot}{O} = C = \overset{\cdot\cdot}{O}\!:$$

Figure 6-3: Formation of carbon dioxide.

Naming Covalent Compounds Made of Two Elements

Binary compounds are compounds made up of only two elements, such as carbon dioxide (CO_2). Chemists use prefixes in the names of binary compounds to indicate the number of atoms of each nonmetal present. Table 6-2 lists the most common prefixes for binary covalent compounds.

Table 6-2	Prefixes for Binary Covalent Compounds		
Number of Atoms	*Prefix*	*Number of Atoms*	*Prefix*
1	mono-	6	hexa-
2	di-	7	hepta-
3	tri-	8	octa-
4	tetra-	9	nona-
5	penta-	10	deca-

In general, the prefix *mono-* is rarely used. Carbon monoxide is one of the few compounds that uses it.

Take a look at the following examples to see how to use the prefixes when naming binary covalent compounds (I've italicized the prefixes):

✔ CO_2: Carbon *di*oxide

✔ P_4O_{10}: *Tetra*phosphorus *deca*oxide (chemists try to avoid putting an *a* and an *o* together with the oxide name, as in dec*a*oxide, so they normally drop the *a* off the prefix)

✔ SO_3: Sulfur *tri*oxide

✔ N_2O_4: *Di*nitrogen *tetr*oxide

This naming system is used only with binary, nonmetal compounds, with one exception: MnO_2 is commonly called *manganese dioxide.*

Writing Covalent Compound Formulas

You can predict the formula of an ionic compound, based on the loss and gain of electrons to reach a noble gas configuration, as I show you in Chapter 5. (For example, if you react Ca with Cl, you can predict the formula of the resulting salt: $CaCl_2$.) However, you really can't make that type of prediction for covalent compounds, because they can combine in many ways, and many different possible covalent compounds may result.

Most of the time, you have to know the formula of the molecule you're studying. But you may have several different types of formulas, and each gives a slightly different amount of information. Oh, joy!

Empirical formulas

An *empirical formula* indicates the different types of elements in a molecule and the lowest whole-number ratio of each kind of atom in the molecule. For example, suppose you have a compound with the empirical formula C_2H_6O. Three different kinds of atoms (C, H, and O) are in the compound, and they're in the lowest whole-number ratio of two carbons to six hydrogens to one oxygen.

Molecular or true formulas

The *molecular formula*, or *true formula*, tells you the kinds of atoms in the compound and the actual number of each atom.

You may determine, for example, that the empirical formula C_2H_6O is actually the molecular formula, too, meaning that there are actually two carbon atoms, six hydrogen atoms, and one oxygen atom in the compound. However, you may instead find that the molecular formula is $C_4H_{12}O_2$, $C_6H_{18}O_3$, $C_8H_{24}O_4$, or another multiple of 2:6:1.

Structural formulas: Dots and dashes

For ionic compounds, the molecular formula is enough to fully identify a compound, but it's not enough to identify covalent compounds. Look at Figure 6-4. Both compounds have the molecular formula of C_2H_6O. That is, both compounds have two carbon atoms, six hydrogen atoms, and one oxygen atom.

However, these are two entirely different compounds with two entirely different sets of properties. The difference is in the way the atoms are bonded, or what's bonded to what. The compound on the left, *dimethyl ether,* is used in some refrigeration units and is highly flammable. The one on the right, *ethyl alcohol,* is the drinking variety of alcohol. Simply knowing the molecular formula isn't enough to distinguish between the two compounds. Can you imagine going into a restaurant, ordering a shot of C_2H_6O, and getting dimethyl ether instead of tequila?

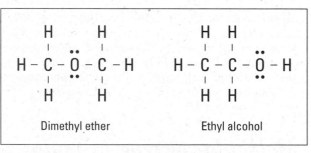

Dimethyl ether   Ethyl alcohol

Figure 6-4: Two possible compounds of C_2H_6O.

Compounds that have the same molecular formula but different structures are called *isomers* of each other. To identify the exact covalent compound, you need its structural formula.

The *structural formula* shows the elements in the compound, the exact number of each atom in the compound, and the bonding pattern for the compound. The electron-dot formula and Lewis formula, which I cover in this section, are common structural formulas.

Basic bonds: Writing the electron-dot and Lewis formulas

The following steps explain how to write the electron-dot formula for a simple molecule — water — and provide some general guidelines to follow:

1. **Write a skeletal structure showing a reasonable bonding pattern using just the element symbols.**

   Often, most atoms are bonded to a single atom. This atom is called the *central atom*. Hydrogen and the halogens are very rarely, if ever, central atoms. Carbon, silicon, nitrogen, phosphorus, oxygen, and sulfur are always good candidates, because they form more than one covalent bond in filling their valence energy level. In the case of water, H_2O, oxygen is the central element and the hydrogen atoms are both bonded to it. The bonding pattern looks like this:

   $$O\begin{matrix} H \\ \\ H \end{matrix}$$

   It doesn't matter where you put the hydrogen atoms around the oxygen. I put the two hydrogen atoms at a 90° angle to each other.

2. **Take all the valence electrons from all the atoms and throw them into an electron pot.**

   Each hydrogen atom has one electron, and the oxygen atom has six valence electrons (VIA family), so you have eight electrons in your electron pot. Those are the electrons you use when making your bonds and completing each atom's octet.

   O H
   H

   electron pot

3. **Use the $N - A = S$ equation to figure the number of covalent bonds in this molecule.**

   In this equation,

   - N **equals the sum of the number of valence electrons needed by each atom.** N has only two possible values — 2 or 8. If the atom is hydrogen, it's 2; if it's anything else, it's 8.

   - A **equals the sum of the number of valence electrons available for each atom.** A is the number of valence electrons in your electron pot. (If you're doing the structure of an ion, you add one electron for every unit of negative charge or subtract one electron for every unit of positive charge.)

   - S **equals the number of electrons shared in the molecule.** And if you divide S by 2, you have the number of covalent bonds in the molecule.

   So in the case of water,

   - $N = 8 + 2(2) = 12$ (eight valence electrons for the oxygen atom, plus two each for the two hydrogen atoms)

   - $A = 6 + 2(1) = 8$ (six valence electrons for the oxygen atom, plus one for each of the two hydrogen atoms)

   - $S = 12 - 8 = 4$ (four electrons shared in water), and S ÷ 2 = 4 ÷ 2 = 2 bonds

   You now know that there are two bonds (two shared pairs of electrons) in water.

4. **Distribute the electrons from your electron pot to account for the bonds.**

   You use four electrons from the eight in the pot, which leaves you with four electrons to distribute later. There has to be at least one bond from your central atom to the atoms surrounding it.

5. **Distribute the rest of the electrons (normally in pairs) so that each atom achieves its full octet of electrons.**

Remember that hydrogen needs only two electrons to fill its valence energy level. In this case, each hydrogen atom has two electrons, but the oxygen atom has only four electrons, so you place the remaining four electrons around the oxygen. This empties your electron pot. Figure 6-5 shows the completed electron-dot formula for water.

$$:\overset{..}{\underset{..}{O}}: H$$
$$H$$

Figure 6-5: Electron-dot formula of H_2O.

Notice that this structural formula shows two types of electrons: *bonding electrons,* the electrons that are shared between two atoms, and *nonbonding electrons,* the electrons that aren't being shared. The last four electrons (two electron pairs) that you put around oxygen aren't being shared, so they're nonbonding electrons.

If you want the Lewis formula, all you have to do is substitute a dash for every bonding pair of electrons in the electron-dot formula. Figure 6-6 shows the Lewis formula for water.

$$:\overset{..}{O} - H$$
$$|$$
$$H$$

Figure 6-6: The Lewis formula for H_2O.

Double bonds: Writing structural formulas for C_2H_4O

Drawing the structural formula for a molecule that contains a double or triple bond can be a bit tricky (see the earlier section "Dealing with multiple bonds"). In those cases, your equations may tell you that you have more covalent bonds that you know what to do with.

For example, here's an example of a structural formula that's a little more complicated — C_2H_4O. The compound has the following framework:

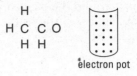

electron pot

Notice that it has not one but two central atoms — the two carbon atoms. You can put 18 valence electrons into the electron pot: four for each carbon atom, one for each hydrogen atom, and six for the oxygen atom.

Now apply the $N - A = S$ equation:

✔ $N = 2(8) + 4(2) + 8 = 32$ (two carbon atoms with eight valence electrons each, plus four hydrogen atoms with two valence electrons each, plus an oxygen atom with eight valence electrons)

✔ $A = 2(4) + 4(1) + 6 = 18$ (four electrons for each of the two carbon atoms, plus one electron for each of the four hydrogen atoms, plus six electrons for the oxygen atom)

✔ $S = 32 - 18 = 14$, and $S \div 2 = 14 \div 2 = 7$ covalent bonds

Put single bonds between the carbon atoms and the hydrogen atoms, between the two carbon atoms, and between the carbon atom and oxygen atom. That's six of your seven bonds.

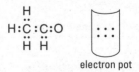

electron pot

There's only one place that the seventh bond can go, and that's between the carbon atom and the oxygen atom. It can't be between a carbon atom and a hydrogen atom, because that would overfill hydrogen's valence energy level. And it can't be between the two carbon atoms, because that would give the carbon on the left ten electrons instead of eight. So there must be a double bond between the carbon atom and the oxygen atom. The four remaining electrons in the pot must be distributed around the oxygen atom, because all the other atoms have reached their octet. Figure 6-7 shows the electron-dot formula.

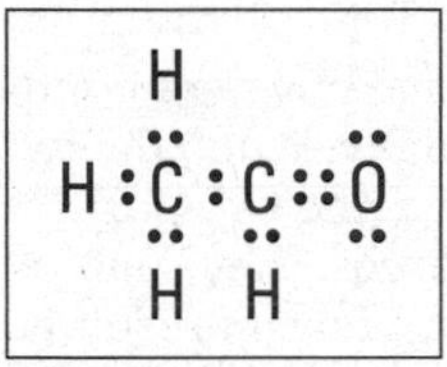

$$H : \overset{\cdot\cdot}{\underset{\cdot\cdot}{C}} : \overset{\cdot\cdot}{\underset{\cdot\cdot}{C}} :: \overset{\cdot\cdot}{\underset{\cdot\cdot}{O}}$$

with H above the first C and H H below.

Figure 6-7: Electron-dot formula of C_2H_4O.

If you convert the bonding pairs to dashes, you have the Lewis formula of C_2H_4O, as in Figure 6-8.

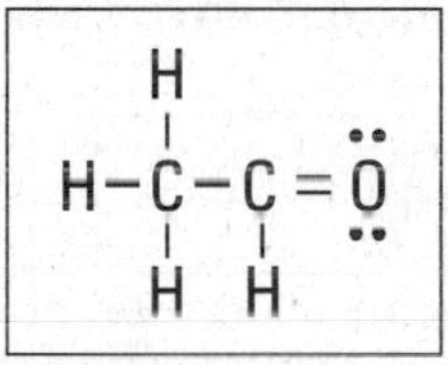

$$H - \underset{H}{\overset{H}{C}} - \underset{H}{C} = \overset{\cdot\cdot}{\underset{\cdot\cdot}{O}}$$

Figure 6-8: The Lewis formula for C_2H_4O.

Grouping atoms with the condensed structural formula

I like the Lewis formula because it enables you to show a lot of information without having to write all those little dots. But it, too, is rather bulky. Sometimes chemists (who are, in general, a lazy lot) use *condensed structural formulas* to show bonding patterns. They may condense the Lewis formula by omitting the nonbonding electrons (dots) and grouping atoms together and/or by omitting certain dashes (covalent bonds). For instance, condensed formulas often group all the hydrogens bonded to a particular carbon atom.

Figure 6-9 shows a couple of condensed formulas for C_2H_4O.

$$CH_3 - CH = O$$

$$CH_3CHO$$

Figure 6-9: Condensed structural formulas for C_2H_4O.

Electronegativities: Which Atoms Have More Pull?

Atoms may share electrons through covalent bonds, but that doesn't mean they share equally. When the two atoms involved in a bond aren't the same, the two positively charged nuclei have different attractive forces; they "pull" on the electron pair to different degrees. The end result is that the electron pair is shifted toward one atom. But the question is, "Which atom does the electron pair shift toward?" Electronegativities provide the answer.

Electronegativity is the strength an atom has to attract a bonding pair of electrons to itself. The larger the electronegativity value, the greater the atom's strength to attract a bonding pair of electrons.

Figure 6-10 shows the electronegativity values of the various elements below each element symbol on the periodic table. Notice that with a few exceptions, the electronegativities increase from left to right in a period and decrease from top to bottom in a family.

Predicting the type of bond

Electronegativities are useful because they give information about what will happen to the bonding pair of electrons when two atoms bond.

A bond in which the electron pair is equally shared is called a *nonpolar covalent bond*. You have a nonpolar covalent bond anytime the two atoms involved in the bond are the same or anytime the difference in the electronegativities of the atoms involved in the bond is very small. For example, consider the Cl_2 molecule. The table in Figure 6-10 shows that chlorine has an electronegativity value of 3.0. Each chlorine atom attracts the bonding electrons with a force of 3.0. Because there's an equal attraction, the bonding electron pair is shared equally between the two chlorine atoms and is located halfway between the two atoms.

Decreasing →

Increasing ↑

1 H 2.1																	
3 Li 1.0	4 Be 1.5											5 B 2.0	6 C 2.5	7 N 3.0	8 O 3.5	9 F 4.0	
11 Na 0.9	12 Mg 1.2											13 Al 1.5	14 Si 1.8	15 P 2.1	16 S 2.5	17 Cl 3.0	
19 K 0.8	20 Ca 1.0	21 Sc 1.3	22 Ti 1.5	23 V 1.6	24 Cr 1.6	25 Mn 1.5	26 Fe 1.8	27 Co 1.9	28 Ni 1.9	29 Cu 1.9	30 Zn 1.6	31 Ga 1.6	32 Ge 1.8	33 As 2.0	34 Se 2.4	35 Br 2.8	
37 Rb 0.8	38 Sr 1.0	39 Y 1.2	40 Zr 1.4	41 Nb 1.6	42 Mo 1.8	43 Tc 1.9	44 Ru 2.2	45 Rh 2.2	46 Pd 2.2	47 Ag 1.9	48 Cd 1.7	49 In 1.7	50 Sn 1.8	51 Sb 1.9	52 Te 2.1	53 I 2.5	
55 Cs 0.7	56 Ba 0.9	57 La 1.1	72 Hf 1.3	73 Ta 1.5	74 W 1.7	75 Re 1.9	76 Os 2.2	77 Ir 2.2	78 Pt 2.2	79 Au 2.4	80 Hg 1.9	81 Tl 1.8	82 Pb 1.9	83 Bi 1.9	84 Po 2.0	85 At 2.2	
87 Fr 0.7	88 Ra 0.9	89 Ac 1.1															

Electronegativities of the Elements

Figure 6-10: Electronegativities of the elements.

A bond in which the electron pair is shifted toward one atom is called a *polar covalent bond*. The atom that more strongly attracts the bonding electron pair is slightly more negative, and the other atom is slightly more positive. The larger the difference in the electronegativities, the more negative and positive the atoms become. Consider hydrogen chloride (HCl). Hydrogen has an electronegativity of 2.1, and chlorine has an electronegativity of 3.0. Because chlorine has a larger electronegativity value, the electron pair that's bonding HCl together shifts toward the chlorine atom.

If the two atoms have extremely different electronegativities, the atoms will probably form ionic, not covalent bonds. For instance, sodium chloride (NaCl) is ionically bonded. An electron has transferred from sodium to chlorine. Sodium has an electronegativity of 1.0, and chlorine has an electronegativity of 3.0. That's an electronegativity difference of 2.0 (3.0 – 1.0), making the bond between the two atoms very, very polar.

The electronegativity difference provides another way of predicting which kind of bond will form between two elements. Here are some guidelines on whether a bond will be polar or nonpolar, covalent or ionic:

Electronegativity Difference	*Type of Bond Formed*
0.0 to 0.2	Nonpolar covalent
0.3 to 1.4	Polar covalent
> 1.5	Ionic

The presence of a polar covalent bond in a molecule can have some pretty dramatic effects on the properties of a molecule, as you see in the next section.

Polar covalent bonding: Creating partial charges

If the two atoms involved in the covalent bond are not the same, the bonding pair of electrons is pulled toward one atom, with that atom taking on a slight (partial) negative charge and the other atom taking on a partial positive charge.

In most cases, the molecule has a positive end and a negative end, called a *dipole* (think of a magnet). Figure 6-11 shows a couple of examples of molecules in which dipoles have formed. (The δ symbol by the charges is the lowercase Greek letter delta, and it refers to a *partial* charge.)

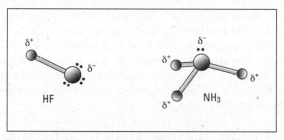

Figure 6-11: Polar covalent bonding in HF and NH₃.

In hydrogen fluoride (HF), the bonding electron pair is pulled much closer to the fluorine atom than to the hydrogen atom, so the fluorine end becomes partially negatively charged and the hydrogen end becomes partially positively charged. The same thing takes place in ammonia (NH₃): The nitrogen has a greater electronegativity than hydrogen, so the bonding pairs of electrons are more attracted to it than to the hydrogen atoms. The nitrogen atom takes on a partial negative charge, and each hydrogen atom takes on a partial positive charge.

The presence of a polar covalent bond explains why some substances act the way they do in a chemical reaction: Because a polar molecule has a positive end and a negative end, it can attract the part of another molecule that has the opposite charge. (See the next section for details.)

In addition, a polar covalent molecule can act as a weak electrolyte because a polar covalent bond allows the substance to act as a conductor. So if a chemist wants a material to act as a good *insulator* (a device used to separate conductors), he or she looks for a material with as weak of a polar covalent bond as possible.

Attracting other molecules: Intermolecular forces

A polar molecule is a *dipole* — with one end having a partial negative charge and the other end having a partial positive charge — so it acts like a magnet. These charged ends can attract other molecules. For instance, the partially negatively charged oxygen atom of one water molecule can attract the partially positively charged hydrogen atom of another water molecule. This attraction between the molecules occurs frequently and is a type of *intermolecular force* (force between different molecules).

Intermolecular forces can be of three different types:

- ✓ **London force (dispersion force):** This very weak type of attraction generally occurs between nonpolar covalent molecules, such as nitrogen (N_2), hydrogen (H_2), or methane (CH_4). It results from the ebb and flow of the electron orbitals, giving a very weak and very brief charge separation around the bond.

- ✓ **Dipole-dipole interaction:** This intermolecular force occurs when the positive end of one dipole molecule is attracted to the negative end of another dipole molecule. It's much stronger than a London force, but it's still pretty weak.

- ✓ **Hydrogen bond:** The third type of interaction is really just an extremely strong dipole-dipole interaction that occurs when a hydrogen atom is bonded to one of three extremely electronegative elements: O, N, or F. These three elements have a very strong attraction for the bonding pair of electrons, so the atoms involved in the bond take on a large amount of partial charge. This bond turns out to be highly polar — and the higher the polarity, the more effective the bond.

When the O, N, or F on one molecule attracts the hydrogen of another molecule, the dipole-dipole interaction is very strong. This strong interaction (only about 5 percent of the strength of an ordinary covalent bond but still very strong for an intermolecular force) is called a *hydrogen bond*. The hydrogen bond is the type of interaction that's present in water.

Chapter 7

Chemical Reactions

*I*n a chemical reaction, substances (elements and/or compounds) are changed into other substances (compounds and/or elements). You can't change one element into another element in a chemical reaction — that happens in nuclear reactions, as I describe in Chapter 4.

A number of clues show that a chemical reaction has taken place — something new is visibly produced, a gas is created, heat is given off or taken in, and so on.

In this chapter, I discuss chemical reactions — how they occur and how to write a balanced chemical equation. I also tell you about chemical equilibrium and explain why chemists often can't get the amount of product out of a reaction that they thought they could. And finally, I discuss the speed of reaction.

Reactants and Products: Reading Chemical Equations

You create a new substance with chemical reactions. The chemical substances that are eventually changed are called the *reactants*, and the new substances that are formed are called the *products*.

Chemical equations show the reactants and products, as well as other factors such as energy changes, catalysts, and so on. With these equations, you use an arrow to indicate that a chemical reaction has taken place. Beneath or above this arrow people sometimes indicate that a specific catalyst is used, or acidic conditions or heat is applied, and so on. In general terms, a chemical reaction follows this format:

Reactants → Products

For example, take a look at the reaction that occurs when you light your natural gas range in order to fry your breakfast eggs. Methane (natural gas) reacts with the oxygen in the atmosphere to produce carbon dioxide and water vapor. (If your burner isn't properly adjusted to give that nice blue flame, you may also get a significant amount of carbon monoxide along with carbon dioxide.) You write the chemical equation that represents this reaction like this:

$$CH_4(g) + 2\,O_2(g) \rightarrow CO_2(g) + 2\,H_2O(g)$$

You can read the equation like this: One molecule of methane gas, $CH_4(g)$, reacts with two molecules of oxygen gas, $O_2(g)$, to form one molecule of carbon dioxide gas, $CO_2(g)$, and two molecules of water vapor, $H_2O(g)$. The 2 in front of the oxygen gas and the 2 in front of the water vapor are called the reaction *coefficients*. They indicate the number of each chemical species that reacts or is formed. I show you how to figure out the value of the coefficients in the section "Balancing Chemical Equations," later in the chapter.

Methane and oxygen are the reactants, and carbon dioxide and water are the products. All the reactants and products are gases (indicated by the *g*'s in parentheses).

In this reaction, all reactants and products are invisible, but you do feel another product of the reaction: heat. The heat being evolved is the clue that tells you a reaction is taking place.

Collision Theory: How Reactions Occur

For a chemical reaction to take place, the reactants must collide. This collision transfers *kinetic energy* (energy of motion) from one substance to the other. The collision between the molecules provides the energy needed to break the necessary bonds so that new bonds can form. In this section, I discuss two criteria for breaking bonds: The reactants have to collide in the right place, and they have to hit with enough energy to break the bonds.

Hitting the right spot

The molecules must collide in the right orientation, or hit at the right spot, in order for the reaction to occur. The place on the molecule where the collision must take place is called the *reactive site*. For example, suppose you have an equation showing molecule A-B reacting with C to form C-A and B, like this:

A-B + C → C-A + B

The way this equation is written, the reaction requires that reactant C collide with A-B on the A end of the molecule. (You know this because the product side shows C hooked up with A — C-A.) If it hits the B end, nothing will happen. The A end of this hypothetical molecule is the reactive site. If C collides at the A end of the molecule, then there's a chance that enough energy can be transferred to break the A-B bond. After the A-B bond is broken, the C-A bond can form. You can show the equation for this reaction process in this way (I show the breaking of the A-B bond and the forming of the C-A bond as "squiggly" bonds):

C~A~B → C-A + B

So for this reaction to occur, there must be a collision between C and A-B at the reactive site.

Note that this example is a simple one. I've assumed that only one collision is needed, making this a one-step reaction. Many reactions are one-step, but many others require several steps in going from reactants to final products. In the process, several compounds may be formed that react with each other to give the final products. These compounds are called *intermediates*. You show them in the reaction *mechanism*, the series of steps that the reaction goes through in going from reactants to products. But in this chapter, I keep it simple and pretty much limit my discussion to one-step reactions.

Adding, releasing, and absorbing energy

REMEMBER

Energy is required to break a bond between atoms. For instance, look at the sample equation A-B + C $\rightarrow$ C-A + B. The collision between C and A-B has to transfer enough energy to break the A-B bond, allowing the C-A bond to form.

Sometimes, even if there is a collision, not enough kinetic energy is available to be transferred — the molecules aren't moving fast enough. You can help the situation somewhat by heating the mixture of reactants. The *temperature* is a measure of the average kinetic energy of the molecules; raising the temperature increases the kinetic energy available to break bonds during collisions.

The energy you have to supply to get a reaction going is called the *activation energy* (E_a) of the reaction. Note that even though you've added energy, the energy of the products isn't always higher than the energy of the reactants — heat may be released during the reaction. This section describes two types of reactions, exothermic and endothermic, in which heat is either released or absorbed.

Exothermic reactions: Releasing heat

In an *exothermic reaction*, heat is given off (released) when you go from reactants to products. The reaction between oxygen and methane as you light a gas stove (from the earlier section "Reactants and Products: Reading Chemical Equations") is a good example of an exothermic reaction.

Even though the reaction gives off heat, you do have to put in a little energy — the activation energy — to get the reaction going. You have to ignite the methane coming out of the burners with a match, lighter, pilot light, or built-in electric igniter.

Imagine that the hypothetical reaction A-B + C → C-A + B is exothermic. The reactants start off at a higher energy state than the products, so energy is released in going from reactants to products. Figure 7-1 shows an energy diagram of this reaction.

In the figure, E_a is the activation energy for the reaction. I show the collision of C and A-B with the breaking of the A-B bond and the forming of the C-A bond at the top of an activation-energy hill. This grouping of reactants at the top of the activation-energy hill is sometimes called the *transition state* of the reaction. This transition state shows what bonds are being broken and what bonds are being made. As I show in Figure 7-1, the difference in the energy level of the reactants and the energy level of the products is the amount of energy (heat) that is released in the reaction.

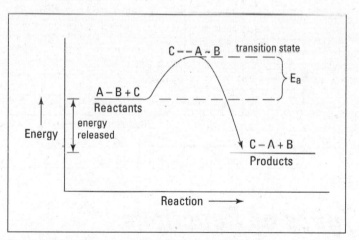

Figure 7-1: Exothermic reaction of A-B + C → C-A + B.

Endothermic reactions: Absorbing heat

Some reactions absorb energy rather than release it. These reactions are called *endothermic* reactions. Cooking involves a lot of endothermic reactions — frying those eggs, for example. You can't just break the shells and let the eggs lie on the pan and then expect the myriad chemical reactions to take place without heating the pan (except when you're outside in Texas during August; there, the sun will heat the pan just fine).

Suppose that the hypothetical reaction A-B + C → C-A + B is endothermic — so the reactants are at a lower energy state than the products. Figure 7-2 shows an energy diagram of this reaction.

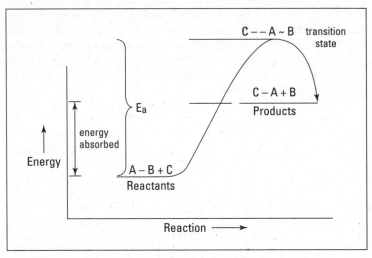

Figure 7-2: Endothermic reaction of A-B + C → C-A + B.

Types of Reactions

Several general types of chemical reactions can occur based upon the identity of the reactants and products and/or which bonds are broken and made. The more common reactions are combination, decomposition, single displacement, double displacement, combustion, and reduction-oxidation (redox) reactions. I describe them all here.

Combination reactions: Coming together

In *combination reactions,* two or more reactants form one product. The reaction of sodium and chlorine to form sodium chloride,

$$2\,Na(s) + Cl_2(g) \rightarrow 2\,NaCl(s)$$

and the burning of coal (carbon) to give carbon dioxide,

$$C(s) + O_2(g) \rightarrow CO_2(g)$$

are examples of combination reactions.

Decomposition reactions: Breaking down

In *decomposition reactions,* a single compound breaks down into two or more simpler substances (elements and/or compounds). Decomposition reactions are the opposite of combination reactions. The decomposition of water into hydrogen and oxygen gases,

$$2\,H_2O(l) \rightarrow 2\,H_2(g) + O_2(g)$$

and the decomposition of hydrogen peroxide to form oxygen gas and water,

$$2\,H_2O_2(l) \rightarrow 2\,H_2O(l) + O_2(g)$$

are examples of decomposition reactions.

Single displacement reactions: Kicking out another element

In *single displacement reactions,* a more active element displaces (kicks out) another less active element from a compound. For example, if you put a piece of zinc metal into a copper (II) sulfate solution, the zinc displaces the copper, as this next equation shows (in case you're wondering, Chapter 5 explains why copper (II) sulfate is named the way it is):

$$Zn(s) + CuSO_4(aq) \rightarrow ZnSO_4(aq) + Cu(s)$$

The notation *(aq)* indicates that the compound is dissolved in water — in an *aqueous* solution.

Because zinc replaces copper in this case, it's said to be more active. If you instead place a piece of copper in a zinc sulfate solution, nothing will happen.

Using the activity series

Table 7-1 shows the activity series of some common metals. Notice that because zinc is more active in the table, it will replace copper, just as the earlier equation shows.

Table 7-1	The Activity Series of Common Metals
Activity	*Metal*
Most active	Alkali and alkaline earth metals
	Al (aluminum)
	Zn (zinc)
	Cr (chromium)
	Fe (iron)
	Ni (nickel)
	Se (selenium)
	Pb (lead)
	Cu (copper)
	Ag (silver)
Least active	Au (gold)

Writing ionic and net-ionic equations

Take a look at the reaction between zinc metal and a copper (II) sulfate solution. I've written this reaction as a molecular equation, showing all species in the neutral form.

$$Zn(s) + CuSO_4(aq) \rightarrow ZnSO_4(aq) + Cu(s)$$

However, these reactions normally occur in an aqueous (water) solution. When you dissolve the ionically bonded $CuSO_4$ in water, it breaks apart into *ions* (atoms or groups of atoms that have an electrical charge due to the loss or gain of electrons). The copper ion has a 2+ charge because it lost two electrons. It's a *cation,* a positively charged ion. The sulfate ion has a 2– charge because it has two extra electrons. It's an *anion,* a negatively charged ion. (Check out Chapter 5 for a more complete discussion of ionic bonding.)

Here's an equation that shows the ions separately. Equations in this form are called *ionic equations* because they show the reaction and production of ions. Notice that the sulfate ion, SO_4^{2-}, doesn't change in the reaction:

$$Zn(s) + Cu^{2+} + SO_4^{2-} \rightarrow Zn^{2+} + SO_4^{2-} + Cu(s)$$

Ions that don't change during the reaction and are found on both sides of the equation in an identical form are called *spectator ions.* Chemists often omit the spectator ions and write the equation showing only those chemical substances that are changed during the reaction. This is called the *net-ionic equation:*

$$Zn(s) + Cu^{2+} \rightarrow Zn^{2+} + Cu(s)$$

Double displacement reactions: Trading places

In single displacement reactions (see the preceding section), only one chemical species is displaced. In *double displacement reactions,* or *metathesis reactions,* two species (normally ions) are displaced. Most of the time, reactions of this type occur in a solution, and either an insoluble solid (in precipitation reactions) or water (in neutralization reactions) will be formed.

Precipitation reactions: Forming solids

The formation of an insoluble solid in a solution is called *precipitation.* For instance, if you mix a solution of potassium chloride and a solution of silver nitrate, a white insoluble solid forms in the resulting solution. Here are the molecular, ionic, and net-ionic equations for this double-displacement reaction:

Molecular: $KCl(aq) + AgNO_3(aq) \rightarrow AgCl(s) + KNO_3(aq)$

Ionic: $K^+ + Cl^- + Ag^+ + NO_3^- \rightarrow AgCl(s) + K^+ + NO_3^-$

Net-ionic: $Cl^- + Ag^+ \rightarrow AgCl(s)$

The white insoluble solid that forms (AgCl) is silver chloride. You can drop out the potassium cation and nitrate anion spectator ions, because they don't change during the reaction and are found on both sides of the equation in an identical form. (See the earlier section "Writing ionic and net-ionic equations" for details on spectator ions.)

To write these equations, you have to know something about the solubility of ionic compounds:

✔ If a compound is soluble, it will not react at all and you can represent it by the appropriate ions or use (aq).

✔ If a compound is insoluble, it will precipitate (form a solid).

Table 7-2 gives the solubilities of selected ionic compounds (see Chapter 5 for info on the names of ionic compounds).

Table 7-2 Solubilities of Selected Ionic Compounds

Water Soluble	Water Insoluble
All chlorides, bromides, iodides . . .	Except those of Ag^+, Pb^{2+}, Hg_2^{2+}
All compound of NH_4^+	Oxides
All compounds of alkali metals	Sulfides
All acetates	Most phosphates
All nitrates	Most hydroxides
All chlorates	
All sulfates . . .	Except $PbSO_4$, $BaSO_4$, and $SrSO_4$

To use Table 7-2, take the cation of one reactant and combine it with the anion of the other reactant (and vice versa), keeping the neutrality of the compounds. This allows you to predict the possible products of the reaction. Then look up the solubilities of the possible products in the table. If the compound is insoluble, it'll precipitate. If it's soluble, it'll remain in solution.

Neutralization reactions: Forming water

Besides precipitation (see the preceding section), the other type of double-displacement reaction is the reaction between an acid and a base. This double-displacement reaction, called a *neutralization reaction,* forms water.

Take a look at the mixing solutions of sulfuric acid (auto battery acid, H_2SO_4) and sodium hydroxide (lye, NaOH). Here are the molecular, ionic, and net-ionic equations for this reaction:

> **Molecular:** $H_2SO_4(aq) + 2 NaOH(aq) \rightarrow Na_2SO_4(aq) + 2 H_2O(l)$
>
> **Ionic:** $2 H^+ + SO_4^{2-} + 2 Na^+ + 2 OH^- \rightarrow 2 Na^+ + SO_4^{2-} + 2 H_2O(l)$
>
> **Net-ionic:** $2 H^+ + 2 OH^- \rightarrow 2 H_2O(l)$ or $H^+ + OH^- \rightarrow H_2O(l)$

To go from the ionic equation to the net-ionic equation, you let the spectator ions (those that don't really react and that appear in an unchanged form on both sides on the arrow) drop out. Then reduce the coefficients in front of the reactants and products down to the lowest common denominator.

Combustion reactions: Burning

Combustion reactions occur when a compound, usually one containing carbon, combines with the oxygen gas in the air. This process is commonly called *burning.* Heat is the most useful product of most combustion reactions.

Here's the equation that represents the burning of propane:

$$C_3H_8(g) + 5 O_2(g) \rightarrow 3 CO_2(g) + 4 H_2O(l)$$

Combustion reactions are also a type of redox reaction.

Redox reactions: Exchanging electrons

Redox reactions, or *reduction-oxidation reactions,* are reactions in which electrons are exchanged. The following three reactions are examples of other types of reactions (such as combination, combustion, and single-replacement reactions), but they're also all redox reactions — they all involve the transfer of electrons from one chemical species to another:

$$2\,Na(s) + Cl_2(g) \rightarrow 2\,NaCl(s)$$

$$C(s) + O_2(g) \rightarrow CO_2(g)$$

$$Zn(s) + CuSO_4(aq) \rightarrow ZnSO_4(aq) + Cu(s)$$

Redox reactions are involved in combustion, rusting, photosynthesis, respiration, the movement of electrons in batteries, and more. I talk about redox reactions in some detail in Chapter 8.

Balancing Chemical Equations

If you carry out a chemical reaction and carefully sum up the masses of all the reactants, and then you compare the sum to the sum of the masses of all the products, you see that they're the same. In fact, a law in chemistry, the *law of conservation of mass*, states, "In an ordinary chemical reaction, matter is neither created nor destroyed." This means that you neither gain nor lose any atoms during the reaction. They may be combined differently, but they're still there.

A chemical equation represents the reaction, and that chemical equation needs to obey the law of conservation of mass. You use that chemical equation to calculate how much of each element you need and how much of each element will be produced. You need to have the same number of each kind of element on both sides of the equation. The equation should balance.

Before you start balancing an equation, you need to know the reactants and the products for that reaction. You can't change the compounds, and you can't change the subscripts, because that would change the compounds. So the only thing you can do to balance the equation is put in *coefficients,* whole numbers in front of the compounds or elements in the equation.

Coefficients tell you how many atoms or molecules you have. For example, if you write *2 H_2O,* it means you have two water molecules:

$$2\,H_2O = H_2O + H_2O$$

Each water molecule is composed of two hydrogen atoms and one oxygen atom. So with 2 H_2O, you have a total of four hydrogen atoms and two oxygen atoms.

In this section, I show you how to balance equations using a method called *balancing by inspection* (or as I call it, "fiddling with coefficients"). You take each atom in turn and balance it by inserting appropriate coefficients on one side or the other. You can balance most simple reactions in this fashion, but one class of reactions is so complex that this method doesn't work well for them: redox reactions. I show you a special method for balancing those equations in Chapter 8.

Balancing the Haber process

My favorite reaction is the *Haber process,* a method for preparing ammonia (NH_3) by reacting nitrogen gas with hydrogen gas:

$$N_2(g) + H_2(g) \rightarrow NH_3(g)$$

This equation shows you what happens in the reaction, but it doesn't show you how much of each element you need to produce the ammonia. To find out how much of each element you need, you have to *balance* the equation — make sure that the number of atoms on the left side of the equation equals the number of atoms on the right. You can't change the subscripts, so you have to put in some coefficients.

In most cases, waiting until the end to balance hydrogen atoms and oxygen atoms is a good idea; balance the other atoms first.

So in this example, you need to balance the nitrogen atoms first. You have two nitrogen atoms on the left side of the arrow (reactant side) and only one nitrogen atom on the right side (product side). To balance the nitrogen atoms, use a coefficient of 2 in front of the ammonia on the right. Now you have two nitrogen atoms on the left and two nitrogen atoms on the right:

$$N_2(g) + H_2(g) \rightarrow 2\ NH_3(g)$$

Next, tackle the hydrogen atoms. You have two hydrogen atoms on the left and six hydrogen atoms on the right (two NH_3 molecules, each with three hydrogen atoms, for a total of six hydrogen atoms). So put a 3 in front of the H_2 on the left, giving you the following:

$$N_2(g) + \mathbf{3}\,H_2(g) \rightarrow 2\,NH_3(g)$$

That should do it. Do a check to be sure: You have two nitrogen atoms on the left and two nitrogen atoms on the right. You have six hydrogen atoms on the left ($3 \times 2 = 6$) and six hydrogen atoms on the right ($2 \times 3 = 6$). The equation is balanced. You can read the equation this way: one nitrogen molecule reacts with three hydrogen molecules to yield two ammonia molecules.

Here's a tidbit for you: This equation would also balance with coefficients of 2, 6, and 4 instead of 1, 3, and 2. In fact, any multiple of 1, 3, and 2 would balance the equation, but chemists have agreed always to show the lowest whole-number ratio (see the discussion of empirical formulas in Chapter 6 for details).

Balancing the burning of butane

Take a look at an equation showing the burning of butane, a hydrocarbon, with excess oxygen available. (This is the reaction that takes place when you light a butane lighter.) The unbalanced reaction is

$$C_4H_{10}(g) + O_2(g) \rightarrow CO_2(g) + H_2O(g)$$

Because waiting until the end to balance hydrogen atoms and oxygen atoms is always a good idea, balance the carbon atoms first. You have four carbon atoms on the left and one carbon atom on the right, so add a coefficient of 4 in front of the carbon dioxide:

$$C_4H_{10}(g) + O_2(g) \rightarrow \mathbf{4}\,CO_2(g) + H_2O(g)$$

Balance the hydrogen atoms next. You have ten hydrogen atoms on the left and two hydrogen atoms on the right, so use a coefficient of 5 in front of the water on the right:

$$C_4H_{10}(g) + O_2(g) \rightarrow 4\,CO_2(g) + \mathbf{5}\,H_2O(g)$$

Now work on balancing the oxygen atoms. You have two oxygen atoms on the left and a total of thirteen oxygen atoms on the right [$(4 \times 2) + (5 \times 1) = 13$]. What can you multiply 2 by to equal 13? How about 6.5?

$$C_4H_{10}(g) + 6.5\ O_2(g) \rightarrow 4\ CO_2(g) + 5\ H_2O(g)$$

But you're not done. You want the lowest *whole-number* ratio of coefficients. Multiply the entire equation by 2 to generate whole numbers:

$$[C_4H_{10}(g) + 6.5\ O_2(g) \rightarrow 4\ CO_2(g) + 5\ H_2O(g)] \times 2$$

Multiply every coefficient by 2 (don't touch the subscripts!) to get

$$2\ C_4H_{10}(g) + 13\ O_2(g) \rightarrow 8\ CO_2(g) + 10\ H_2O(g)$$

If you check the atom count on both sides of the equation, you find that the equation is balanced, and the coefficients are in the lowest whole-number ratio.

 After balancing an equation, make sure that the same number of each atom is on both sides and that the coefficients are in the lowest whole-number ratio.

Knowing Chemical Equilibrium Backwards and Forwards

A *dynamic chemical equilibrium* is established when two exactly opposite chemical reactions are occurring at the same place, at the same time, with the same rates (speed) of reaction. I call this example a *dynamic* chemical equilibrium, because when the reactions reach equilibrium, things don't just stop — the reactants still react to form the products, and those substances react to form the original reactants.

In this section, I explain how reactions reach equilibrium. I also introduce the equilibrium constant, which helps you find out how much product and reactant you have when the reaction is at equilibrium.

Matching rates of change in the Haber process

My favorite reaction is the Haber process, the synthesis of ammonia from nitrogen and hydrogen gases. After balancing the reaction (see the section "Balancing the Haber process," earlier in this chapter), you end up with

$$N_2(g) + 3 H_2(g) \rightarrow 2 NH_3(g)$$

Written this way, the reaction says that hydrogen and nitrogen react to form ammonia — and this keeps on happening until you use up one or both of the reactants. But this isn't quite true.

If this reaction occurs in a closed container (which it has to, with everything being gases), then the nitrogen and hydrogen react and ammonia is formed — but some of the ammonia soon starts to decompose into nitrogen and hydrogen, like this:

$$2 NH_3(g) \rightarrow N_2(g) + 3 H_2(g)$$

In the container, then, you actually have *two* exactly opposite reactions occurring — nitrogen and hydrogen combine to give ammonia, and ammonia decomposes to give nitrogen and hydrogen.

Instead of showing the two separate reactions, you can show one reaction and use a double arrow like this:

$$N_2(g) + 3 H_2(g) \leftrightarrow 2 NH_3(g) + heat$$

You put the nitrogen and hydrogen on the left because that's what you initially put into the reaction container. The reaction is exothermic (it gives off heat), so I'm showing heat as a product of the reaction on the right.

Now these two reactions occur at different speeds, but sooner or later, the two speeds become the same, and the relative amounts of nitrogen, hydrogen, and ammonia become constant. This is an example of a chemical equilibrium. At any given time in the Haber process, you have nitrogen and hydrogen reacting to form ammonia and ammonia decomposing to form nitrogen and hydrogen. When the system reaches equilibrium, the amounts of all chemical species become *constant* but not necessarily the same.

Constants: Comparing amounts of products and reactants

Sometimes there's a lot of product (chemical species on the right-hand side of the double arrow) when the reaction reaches equilibrium, and sometimes there's very little. You can tell the relative amounts of reactants and products at equilibrium if you know the equilibrium constant for the reaction. Look at a hypothetical equilibrium reaction:

$$aA + bB \leftrightarrow cC + dD$$

The capital letters stand for the chemical species, and the small letters represent the coefficients in the balanced chemical equation. The *equilibrium constant* (represented as K_{eq}) is mathematically defined as

$$K_{eq} = \frac{[C]^c [D]^d}{[A]^a [B]^b}$$

The numerator contains the product of the two chemical species on the right-hand side of the equation, with each chemical species raised to the power of its coefficient in the balanced chemical equation. The denominator is the same, but you use the chemical species on the left-hand side of the equation. (It's not important right now, but those brackets stand for the *molar concentration.* You can find out what that is in Chapter 10.) Note that sometimes chemists use the K_c notation instead of the K_{eq} form.

The numerical value of the equilibrium constant gives you a clue about the relative amounts of products and reactants. The larger the value of the equilibrium constant (K_{eq}), the more products are present at equilibrium. If, for example, you have a reaction that has an equilibrium constant of 0.001 at room temperature and 0.1 at 100°C, you can say that you'll have much more product at the higher temperature.

Now I happen to know that the K_{eq} for the *Haber process* (the ammonia synthesis) is 3.5×10^8 at room temperature. This large value indicates that, at equilibrium, there's a lot of ammonia produced from the nitrogen and hydrogen, but there's still hydrogen and nitrogen left at equilibrium.

Le Chatelier's Principle: Getting More (or Less) Product

If you're, say, an industrial chemist, you want as much of the reactants as possible to be converted to product. You'd like the reaction to go to completion (meaning you'd like the reactants to keep creating the product until they're all used up), but that doesn't happen for an equilibrium reaction. But it would be nice if you could, in some way, manipulate the system to get a little bit more product formed. There is such a way — through Le Chatelier's Principle.

A French chemist, Henri Le Chatelier, discovered that if you apply a change of condition (called *stress*) to a chemical system that's at equilibrium, the reaction will return to equilibrium by shifting in such a way as to counteract the change (the stress). This is called *Le Chatelier's Principle*.

You can stress an equilibrium system in three ways:

✔ Change the concentration of a reactant or product.

✔ Change the temperature.

✔ Change the pressure on a system that contains gases.

In this section, you see how this applies to the Haber process:

$$N_2(g) + 3 H_2(g) \leftrightarrow 2 NH_3(g) + heat$$

Changing the concentration

In general, if you add more of a reactant or product to an equilibrium system, the reaction will shift to the other side to use it up. If you remove some reactant or product, the reaction shifts to that side in order to replace it.

Suppose that you have the ammonia system at equilibrium, and you then put in some more nitrogen gas. To reestablish the equilibrium, the reaction shifts from the left to the right, using up some nitrogen and hydrogen, and forming more ammonia and heat.

The equilibrium has been reestablished. You have less hydrogen and more nitrogen, ammonia, and heat than you had before you added the additional nitrogen. The same thing would happen if you had a way of removing ammonia as it was formed. The right-hand side of the teeter-totter would again be lighter, and weight would be shifted to the right in order to reestablish the equilibrium. Again, more ammonia would be formed.

Changing the temperature

In general, heating a reaction causes it to shift to the endothermic (heat-absorbing) side. (If you have an exothermic reaction where heat is produced on the right side, then the left side is the endothermic side.) Cooling a reaction mixture causes the equilibrium to shift to the exothermic (heat-releasing) side.

Suppose that you heat the reaction mixture of nitrogen and hydrogen. You know that the reaction is exothermic — heat is given off, showing up on the right-hand side of the equation. So if you heat the reaction mixture, the reaction shifts to the left to use up the extra heat and reestablish the equilibrium. This shift uses up ammonia and produces more nitrogen and hydrogen. And as the reaction shifts, the amount of heat also decreases, lowering the temperature of the reaction mixture.

Changing the pressure

Changing the pressure affects the equilibrium only if there are reactants and/or products that are gases. In general, increasing the pressure on an equilibrium mixture causes the reaction to shift to the side containing the fewest number of gas molecules.

In the Haber process, all species are gases, so you do see a pressure effect. Think about the sealed container where your ammonia reaction is occurring. (The reaction has to occur in a sealed container because everything is a gas.) You have nitrogen, hydrogen, and ammonia gases inside. There is pressure in the sealed container, and that pressure is due to the gas molecules hitting the inside walls of the container.

Now suppose that the system is at equilibrium, and you want to increase the pressure. You can do so by making the container smaller (with a piston type of arrangement) or by putting in nonreactive gas, such as neon. You get more collisions on the inside walls of the container, and therefore, you have more pressure. Increasing the pressure stresses the equilibrium; to remove that stress and reestablish the equilibrium, the pressure must be reduced.

Take another look at the Haber reaction and look for some clues on how this may happen.

$$N_2(g) + 3\,H_2(g) \leftrightarrow 2\,NH_3(g)$$

Every time the forward (left-to-right) reaction takes place, four molecules of gas (one nitrogen and three hydrogen) form two molecules of ammonia gas. This reaction reduces the number of molecules of gas in the container. The reverse reaction (right-to-left) takes two ammonia gas molecules and makes four gas molecules (nitrogen and hydrogen). This reaction increases the number of gas molecules in the container.

The equilibrium has been stressed by an increase in pressure; reducing the pressure will relieve the stress. Reducing the number of gas molecules in the container will reduce the pressure (fewer collisions on the inside walls of the container), so the forward (left-to-right) reaction is favored because four gas molecules are consumed and only two are formed. As a result of the forward reaction, more ammonia is produced!

Chemical Kinetics: Changing Reaction Speeds

Kinetics is the study of the speed of a reaction. Some reactions are fast; others are slow. Sometimes chemists want to speed the slow ones up and slow the fast ones down. For instance, if you're a chemist who wants to make ammonia from a given amount of hydrogen and nitrogen, you want to produce ammonia *as fast as possible*.

Several factors affect the speed of a reaction:

- ✔ **Complexity of the reactants:** In general, the reaction rate is slower when the reactants are large and complex molecules. For a reaction to occur, the reactants have to collide at the *reactive site* of the molecule. The larger and more complex the reactant molecules are, the less chance of a collision at the reactive site. Sometimes, in very complex molecules, other parts of the molecule totally block off the reactive site, so no reaction occurs. There may be a lot of collisions, but only the ones that occur at the reactive site have any chance of leading to chemical reaction.

- ✔ **Particle size of the reactants:** Reaction depends on collisions. The more surface area on which collisions can occur, the faster the reaction. For instance, you can hold a burning match to a large chunk of coal and nothing will happen. But if you take that same piece of coal, grind it up very, very fine, throw it up into the air, and strike a match, you'll get an explosion because of the increased surface area of the coal.

- ✔ **Concentration of the reactants:** Increasing the number of collisions speeds up the reaction rate. The more reactant molecules that are colliding, the faster the reaction will be. For example, a wood splint burns okay in air (20 percent oxygen), but it burns *much* faster in pure oxygen.

  In most simple cases, increasing the concentration of the reactants increases the speed of the reaction. However, if the reaction is complex and it has a complex *mechanism* (series of steps in the reaction), this may not be the case.

Determining the concentration effect on the rate of reaction can give you clues as to which reactant is involved in the step of the mechanism that determines the reaction speed. (You can then use this information to help figure out the reaction mechanism.) You can do this by running the reaction at several different concentrations and observing the effect on the rate of reaction. If, for example, changing the concentration of one reactant has no effect on the rate of reaction, then you know that reactant is not involved in the slowest step (the rate-determining step) in the mechanism.

✔ **Pressure of gaseous reactants:** The pressure of gaseous reactants has basically the same effect as concentration. The higher the reactant pressure, the faster the reaction rate. This is due to (you guessed it!) the increased number of collisions. But if there's a complex mechanism involved, changing the pressure may not have the result you expect.

✔ **Temperature:** In most cases, increasing the temperature causes the reaction rate to increase. In organic chemistry, a general rule says that increasing the temperature 10°C will cause the reaction rate to double.

But why is this true? Part of the answer is an increased number of collisions, because increasing the temperature causes the molecules to move faster. But this is only part of the story. Increasing the temperature also increases the average kinetic energy of the molecules. Increasing the temperature not only increases the number of collisions but also increases the number of collisions that are effective — that transfer enough energy to cause a reaction to take place.

✔ **Catalysts:** *Catalysts* are substances that increase the reaction rate without themselves being changed at the end of the reaction. They increase the reaction rate by lowering the reaction's *activation energy* (the energy required to start the reaction). Look back at Figure 7-1, for example. If the activation-energy hill were lower, it'd be easier for the reaction to occur and the reaction rate would be faster. I discuss catalysts in the next section.

Seeing How Catalysts Speed Up Reactions

Catalysts speed up reactions by lowering the activation energy of a reaction. They do this in one of two ways:

✔ Providing a surface and orientation that makes a reactant more likely to hit the right part of another reactant to break/make a bond

✔ Providing an alternative *mechanism* (series of steps for the reaction to go through) that has a lower activation energy

Heterogeneous catalysis: Giving reactants a better target

A heterogeneous catalyst ties one molecule to a surface, putting the molecule in an orientation that makes another reactant more likely to hit the reactive site. Suppose, for instance, you have the following generalized reaction (which I introduce earlier in "Collision Theory: How Reactions Occur"):

C~A~B → C-A + B

Reactant C must hit the reactive site on the A end of molecule A-B in order to break the A-B bond and form the C-A bond shown in the equation. The probability of the collision occurring in the proper orientation is pretty much driven by chance. The reactants are moving around, running into each other, and sooner or later the collision may occur at the reactive site. But what would happen if you could tie the A-B molecule down with the A end exposed? It'd be much easier and more probable for C to hit A with this scenario.

The catalyst that does this lying down is called *heterogeneous* because it's in a different phase than the reactants. This catalyst is commonly a finely divided solid metal or metal oxide, and the reactants are gases or in solution. This heterogeneous catalyst tends to attract one part of a reactant molecule due to rather complex interactions that aren't fully understood. After the reaction takes place, the forces that bound the B part of the molecule to the surface of the catalyst are no longer there. So B can drift off, and the catalyst is ready to do it again.

Most people sit very close to a heterogeneous catalyst every day — the catalytic converter in an automobile. It contains finely divided platinum and/or palladium metal and speeds up the reaction that causes harmful gases from gasoline (such as carbon monoxide and unburned hydrocarbons) to decompose into mostly harmless products (such as water and carbon dioxide).

Homogeneous catalysis: Offering an easier path

The second type of catalyst is a *homogeneous catalyst* — one that's in the same phase as the reactants. It provides an alternative mechanism, or reaction pathway, that has a lower activation energy than the original reaction.

For an example, check out the decomposition reaction of hydrogen peroxide:

$$2 H_2O_2(l) \rightarrow 2 H_2O(l) + O_2(g)$$

This is a slow reaction, especially if you keep the hydrogen peroxide cool in a dark bottle. The hydrogen peroxide in that bottle in your medicine cabinet may take years to decompose. But if you put a little bit of a solution containing the ferric ion in the bottle, the reaction will be much faster, even though it'll be a two-step mechanism instead of a one-step mechanism:

Step 1: $2 Fe^{3+} + H_2O_2(l) \rightarrow 2 Fe^{2+} + O_2(g) + 2 H^+$

Step 2: $2 Fe^{2+} + H_2O_2(l) + 2 H^+ \rightarrow 2 Fe^{3+} + 2 H_2O(l)$

If you add the two preceding reactions together and cancel the species that are identical on both sides, you get the original, uncatalyzed reaction (species to be cancelled are bolded):

2 Fe^{3+} + $H_2O_2(l)$ + **2 Fe^{2+}** + $\rightarrow$ **2 Fe^{2+}** + $O_2(g)$ + **2 H$^+$** + $H_2O_2(l)$ + **2 H$^+$**   **2 Fe^{3+}** + 2 $H_2O(l)$

$$2 H_2O_2(l) \rightarrow 2 H_2O(l) + O_2(g)$$

The ferric ion catalyst was changed in the first step and then changed back in the second step. This two-step catalyzed pathway has a lower activation energy and is faster.

Chapter 8

Electrochemistry: Using Electrons

● ●

In This Chapter

▶ Finding out about redox reactions

▶ Balancing redox equations

▶ Taking a look at electrochemical cells

● ●

Combustion is a redox reaction. So are respiration, photo-synthesis, and many other biochemical processes people depend on for life. In this chapter, I explain redox reactions, go through the balancing of this type of equation, and then show you some applications of redox reactions in an area of chemistry called electrochemistry. Electrochemistry is an area of chemistry in which we use chemical reactions to produce electrons (electricity) or use electrons (electricity) to cause a desired chemical reaction to take place.

Transferring Electrons with Redox Reactions

Redox reactions — reactions in which there's a simultaneous transfer of electrons from one chemical species (chemical entity such as an atom or molecule) to another — are really composed of two different reactions:

 ✔ **Oxidation:** A loss of electrons

 ✔ **Reduction:** A gain of electrons

These reactions are coupled, because the electrons that are lost in the oxidation reaction are the same electrons that are gained in the reduction reaction. In fact, these two reactions (reduction and oxidation) are commonly called *half-reactions,* because you need these two halves to make a whole reaction, and the overall reaction is called a *redox (redu*ction/*oxi*dation) reaction. In Chapter 7, I describe a redox reaction that occurs between zinc metal and the cupric (copper II, Cu^{2+}) ion. The zinc metal loses electrons and the copper II ion gains them.

Oxidation

You can use three definitions for oxidation:

- ✔ The loss of electrons
- ✔ The gain of oxygen
- ✔ The loss of hydrogen

Because I typically deal with electrochemical cells, I normally use the definition that describes the loss of the electrons. The other definitions are useful in processes such as combustion and photosynthesis.

Loss of electrons

One way to define oxidation is with the reaction in which a chemical substance loses electrons in going from reactant to product. For example, when sodium metal reacts with chlorine gas to form sodium chloride (NaCl), the sodium metal loses an electron, which chlorine then gains. The following equation shows sodium losing the electron:

$$Na(s) \rightarrow Na^+ + e^-$$

When it loses the electron, chemists say that the sodium metal has been oxidized to the sodium cation. (A *cation* is an ion with a positive charge due to the loss of electrons — see Chapter 5.)

Reactions of this type are quite common in *electrochemical reactions,* reactions that produce or use electricity.

Gain of oxygen

In certain oxidation reactions, it's obvious that oxygen has been gained in going from reactant to product. Reactions where the gain of oxygen is more obvious than the gain of electrons include combustion reactions *(burning)* and the *rusting* of iron. Here are two examples:

$$C(s) + O_2(g) \rightarrow CO_2(g) \qquad \text{(burning of coal)}$$

$$2\,Fe(s) + 3\,O_2(g) \rightarrow 2\,Fe_2O_3(s) \qquad \text{(rusting of iron)}$$

In these cases, chemists say that the carbon and the iron metal have been oxidized to carbon dioxide and rust, respectively.

Loss of hydrogen

In other reactions, you can best see oxidation as the loss of hydrogen. Methyl alcohol (wood alcohol) can be oxidized to formaldehyde:

$$CH_3OH(l) \rightarrow CH_2O(l) + H_2(g)$$

In going from methanol to formaldehyde, the compound goes from having four hydrogen atoms to having two hydrogen atoms.

Reduction

You can use three definitions to describe reduction:

✔ The gain of electrons

✔ The loss of oxygen

✔ The gain of hydrogen

Gain of electrons

Chemists often see reduction as the gain of electrons. In the process of electroplating silver onto a teapot, for example, the silver cation is reduced to silver metal by the gain of an electron. The following equation shows the silver cation's gaining the electron:

$$Ag^+ + e^- \rightarrow Ag$$

When it gains the electron, chemists say that the silver cation has been reduced to silver metal.

Loss of oxygen

In some reactions, seeing reduction as the loss of oxygen in going from reactant to product is easy. For example, a reaction with carbon monoxide in a blast furnace reduces iron ore (primarily rust, Fe_2O_3) is to iron metal:

$$Fe_2O_3(s) + 3\ CO(g) \rightarrow 2\ Fe(s) + 3\ CO_2(g)$$

The iron has lost oxygen, so chemists say that the iron ion has been reduced to iron metal.

Gain of hydrogen

In certain cases, you can describe a reduction as the gain of hydrogen atoms in going from reactant to product. For example, carbon monoxide and hydrogen gas can be reduced to methyl alcohol:

$$CO(g) + 2\ H_2(g) \rightarrow CH_3OH(l)$$

In this reduction process, the CO has gained the hydrogen atoms.

One's loss is the other's gain

Neither oxidation nor reduction can take place without the other. When those electrons are lost, something has to gain them. Consider, for example, the *net-ionic equation* (the equation showing just the chemical substances that are changed during a reaction — see Chapter 7) for a reaction with zinc metal and an aqueous copper(II) sulfate solution:

$$Zn(s) + Cu^{2+} \rightarrow Zn^{2+} + Cu$$

This overall reaction is really composed of two half-reactions:

$Zn(s) \rightarrow Zn^{2+} + 2e^-$ (oxidation half-reaction — the loss of electrons)

$Cu^{2+} + 2e^- \rightarrow Cu(s)$ (reduction half-reaction — the gain of electrons)

To help yourself remember which reaction is oxidation and which is reduction in terms of electrons, memorize the phrase "LEO goes GER" (Lose Electrons Oxidation; Gain Electrons Reduction).

Zinc loses two electrons; the copper(II) cation gains those same two electrons. Zn is being oxidized. But without Cu^{2+} present, nothing will happen. That copper cation is the *oxidizing agent*. It's a necessary agent for the oxidation process to proceed. The oxidizing agent accepts the electrons from the chemical species that's being oxidized.

Cu^{2+} is reduced as it gains electrons. The species that furnishes the electrons is the *reducing agent*. In this case, the reducing agent is zinc metal.

The oxidizing agent is the species that's being reduced, and the reducing agent is the species that's being oxidized. Both the oxidizing and reducing agents are on the left (reactant) side of the redox equation.

Oxidation numbers

Oxidation numbers are bookkeeping numbers. They allow chemists to do things such as balance redox equations. Oxidation numbers are positive or negative numbers, but don't confuse them with charges on ions or valences. Chemists assign oxidation numbers to elements using these rules:

- ✔ **For free elements:** The oxidation number of an element in its free (uncombined) state is zero (for example, Al(s) or Zn(s)). This is also true for elements found in nature as *diatomic* (two-atom) elements (H_2, O_2, N_2, F_2, Cl_2, Br_2, or I_2) and for sulfur, found as S_8.

- ✔ **For single-atom ions:** The oxidation number of a *monatomic* (one-atom) ion is the same as the charge on the ion (for example, $Na^+ = +1$, $S^{2-} = -2$).

- ✔ **For compounds:** The sum of all oxidation numbers in a neutral compound is zero. The sum of all oxidation numbers in a *polyatomic* (many-atom) ion is equal to the charge on the ion. This rule often allows chemists to calculate the oxidation number of an atom that may have multiple oxidation states, if the other atoms in the ion have known oxidation numbers. (See Chapter 6 for examples of atoms with multiple oxidation states.)

✔ **For alkali metals and alkaline earth metals in compounds:** The oxidation number of an alkali metal (IA family) in a compound is +1; the oxidation number of an alkaline earth metal (IIA family) in a compound is +2.

✔ **For oxygen in compounds:** The oxidation number of oxygen in a compound is usually –2. If, however, the oxygen is in a class of compounds called *peroxides* (for example, hydrogen peroxide, or H_2O_2), then the oxygen has an oxidation number of –1. If the oxygen is bonded to fluorine, the number is +1.

✔ **For hydrogen in compounds:** The oxidation state of hydrogen in a compound is usually +1. If the hydrogen is part of a *binary metal hydride* (compound of hydrogen and some metal), then the oxidation state of hydrogen is –1.

✔ **For halogens:** The oxidation number of fluorine is always –1. Chlorine, bromine, and iodine usually have an oxidation number of –1, unless they're in combination with an oxygen or fluorine. (For example, in ClO^-, the oxidation number of oxygen is –2 and the oxidation number of chlorine is +1; remember that the sum of all the oxidation numbers in ClO^- has to equal –1.)

These rules give you another way to define oxidation and reduction — in terms of oxidation numbers. For example, consider this reaction, which shows oxidation by the loss of electrons:

$$Zn(s) \rightarrow Zn^{2+} + 2e^-$$

Notice that the zinc metal (the reactant) has an oxidation number of zero (the first rule), and the zinc cation (the product) has an oxidation number of +2 (the second rule). In general, you can say that a substance is oxidized when there's an *increase* in its oxidation number.

Reduction works the same way. Consider this reaction:

$$Cu^{2+} + 2e^- \rightarrow Cu(s)$$

The copper is going from an oxidation number of +2 to zero. A substance is reduced if there's a *decrease* in its oxidation number.

Balancing Redox Equations

Redox equations are often so complex that the inspection method (the fiddling-with-coefficients method) of balancing chemical equations doesn't work well with them. (See Chapter 7 for a discussion of this balancing method.) So chemists developed other methods of balancing redox equations, such as the *ion electron* (half-reaction) method.

Here's an overview of how it works: You convert the unbalanced redox equation to the ionic equation and then break it down into two half-reactions — oxidation and reduction. Balance each of these half-reactions separately and then combine them to give the balanced ionic equation. Finally, put the spectator ions into the balanced ionic equation, converting the reaction back to the molecular form. (For a discussion of molecular, ionic, and net-ionic equations, see Chapter 7.)

To successfully balance redox equations with the ion electron method, you need to follow the steps precisely and in order. Here's what to do:

1. **Convert the unbalanced redox reaction to the ionic form.**

   Suppose you want to balance this redox equation:

   $$Cu(s) + HNO_3(aq) \rightarrow Cu(NO_3)_2(aq) + NO(g) + H_2O(l)$$

   In this reaction, you show the nitric acid in the ionic form, because it's a strong acid (for a discussion of strong acids, see Chapter 11). Copper(II) nitrate is soluble (indicated by (aq)), so show it in its ionic form (see Chapter 7). Because NO(g) and water are molecular compounds, they remain in the molecular form:

   $$Cu(s) + H^+ + NO_3^- \rightarrow Cu^{2+} + 2\, NO_3^- + NO(g) + H_2O(l)$$

2. **If necessary, assign oxidation numbers; then write two half-reactions (oxidation and reduction) that show the chemical species that have had their oxidation numbers changed.**

   In some cases, telling what's been oxidized and reduced is easy; in other cases, it isn't. Start by going through the example reaction and assigning oxidation numbers (see the earlier section "Oxidation numbers" for details):

   $$Cu(s) + H^+ + NO_3^- \rightarrow Cu^{2+} + 2\ NO_3^- + NO(g) + H_2O(l)$$

   $$0 \qquad +1\ +5\ (-2)3 \quad +2 \quad +5\ (-2)3 \quad +2\ -2 \quad (+1)2\ -2$$

   To write your half-reactions, look closely for places where the oxidation numbers have changed, and write down those chemical species. Copper changes its oxidation number (from 0 to 2) and so does nitrogen (from −5 to +2), so your unbalanced half-reactions are

   $$Cu(s) \rightarrow Cu^{2+}$$

   $$NO_3^- \rightarrow NO$$

3. **Balance all atoms, with the exception of oxygen and hydrogen.**

   Starting with the two unbalanced half-reactions above, you can balance atoms other than oxygen and hydrogen by *inspection* — fiddling with the coefficients. (You can't change subscripts; you can only add coefficients.) In this case, both the copper and nitrogen atoms already balance, with one each on both sides:

   $$Cu(s) \rightarrow Cu^{2+}$$

   $$NO_3^- \rightarrow NO$$

4. **Balance the oxygen atoms.**

   How you balance these atoms depends on whether you're dealing with acid or basic solutions:

   - In acidic solutions, take the number of oxygen atoms needed and add that same number of water molecules to the side that needs oxygen.

   - In basic solutions, add two OH⁻ to the side that needs oxygen for every oxygen atom that is needed. Then, to the other side of the equation, add half as many water molecules as OH^- anions used.

An acidic solution shows some acid or H^+; a basic solution has an OH^- present. The example equation is in acidic conditions (nitric acid, HNO_3, which is $H^+ + NO_3^-$ in ionic form). You don't have to do anything on the half-reaction involving the copper, because no oxygen atoms are present. But you do need to balance the oxygen atoms in the second half-reaction:

$$Cu(s) \rightarrow Cu^{2+}$$

$$NO_3^- \rightarrow NO + 2\,H_2O$$

5. **Balance the hydrogen atoms.**

Again, how you balance these atoms depends on whether you're dealing with acid or basic solutions:

- In acidic solutions, take the number of hydrogen atoms needed and add that same number of H^+ to the side that needs hydrogen.

- In basic solutions, add one water molecule to the side that needs hydrogen for every hydrogen atom that's needed. Then, to the other side of the equation, add as many OH^- anions as water molecules used.

The example equation is in acidic conditions. You need to balance the hydrogen atoms in the second half-reaction:

$$Cu(s) \rightarrow Cu^{2+}$$

$$4\,H^+ + NO_3^- \rightarrow NO + 2\,H_2O$$

6. **Balance the ionic charge on each half-reaction by adding electrons.**

$$Cu(s) \rightarrow Cu^{2+} + 2\,e^- \text{ (oxidation)}$$

$$3\,e^- + 4\,H^+ + NO_3^- \rightarrow NO + 2\,H_2O \text{ (reduction)}$$

The electrons should end up on opposite sides of the equation in the two half-reactions. Remember that you're using ionic charge, not oxidation numbers.

7. **Balance electron loss with electron gain between the two half-reactions.**

The electrons that are lost in the oxidation half-reaction are the same electrons that are gained in the reduction half-reaction, so the number of electrons lost and gained must be the same. But Step 6 shows a loss of two electrons and a gain of three. So you must adjust the numbers using appropriate multipliers for both half-reactions. In this case, you have to find the lowest common multiple of 2 and 3. It's 6, so multiply the first half-reaction by 3 and the second half-reaction by 2.

$$3 \times [Cu(s) \rightarrow Cu^{2+} + 2\ e^-] = 3\ Cu(s) \rightarrow 3\ Cu^{2+} + 6\ e^-$$

$$2 \times [3\ e^- + 4\ H^+ + NO_3^- \rightarrow NO + 2\ H_2O] =$$
$$6\ e^- + 8\ H^+ + 2\ NO_3^- \rightarrow 2\ NO + 4\ H_2O$$

8. **Add the two half-reactions together and cancel anything common to both sides.**

The electrons should always cancel (the number of electrons should be the same on both sides).

$$3\ Cu + 6\ e^- + 8\ H^+ + 2\ NO_3^- \rightarrow 3\ Cu^{2+} + 6\ e^- + 2\ NO + 4\ H_2O$$

9. **Convert the equation back to the molecular form by adding the spectator ions.**

If it's necessary to add spectator ions (ions not involved in the reaction, but there to ensure electrical neutrality – Chapter 7) to one side of the equation in order to convert it back to the molecular equation, add the same number to the other side of the equation. For example, there are eight H^+ on the left side of the equation. In the original equation, the H^+ was in the molecular form of HNO_3. You need to add the NO_3^- spectator ions back to it. You already have 2 on the left, so simply add 6 more. You then add 6 NO_3^- to the right-hand side to keep things balanced. Those are the spectator ions that you need for the Cu^{2+} cation to convert it back to the molecular form that you want.

$$3\ Cu(s) + 8\ HNO_3(aq) \rightarrow 3\ Cu(NO_3)_2(aq) +$$
$$2\ NO(g) + 4\ H_2O(l)$$

10. **Check to make sure that all the atoms are balanced, all the charges are balanced (if working with an ionic equation at the beginning), and all the coefficients are in the lowest whole-number ratio.**

That's how it's done. Reactions that take place in bases are just as easy, as long as you follow the rules.

Exploring Electrochemical Cells

Redox reactions sometimes involve *direct* electron transfer, in which one substance immediately picks up the electrons another has lost. For instance, if you put a piece of zinc metal into a copper(II) sulfate solution, zinc gives up two electrons (becomes oxidized) to the Cu^{2+} ion that accepts the electrons (reducing it to copper metal). The copper metal begins spontaneously plating out on the surface of the zinc. The equation for the reaction is

$$Zn(s) + Cu^{2+} \rightarrow Zn^{2+} + Cu$$

But if you separate those two half-reactions so that when the zinc is oxidized, the electrons it releases are forced to travel through a wire to get to the Cu^{2+}, you get something useful: a *galvanic* or *voltaic cell,* a redox reaction that produces electricity. In this section, I show you how that Zn/Cu^{2+} reaction may be separated out so that you have an *indirect* electron transfer and can produce some useable electricity. I also show you how *electrolytic cells* do the reverse, using electricity to cause a redox reaction. Finally, you see how rechargeable batteries both generate electricity and cause chemical reactions.

Galvanic cells: Getting electricity from chemical reactions

Galvanic cells use redox reactions to produce electricity. These cells are commonly called batteries, but sometimes this name is somewhat incorrect, because a *battery* is composed of two or more cells connected together. You put a battery in your car, but you put a cell into your flashlight.

A Daniell cell is a type of galvanic cell that uses the Zn/Cu^{2+} reaction to produce electricity. In the Daniell cell, a piece of zinc metal is placed in a solution of zinc sulfate in one container, and a piece of copper metal is placed in a solution of copper(II) sulfate in another container. These strips of metal are called the cell's *electrodes*. They act as a terminal, or a holding place, for electrons.

A wire connects the electrodes, but nothing happens until you put a salt bridge between the two containers. The *salt bridge,* normally a U-shaped hollow tube filled with a concentrated salt solution, provides a way for ions to move from one container to the other to keep the solutions electrically neutral. It's like running only one wire up to a ceiling light; the light won't work unless you put in a second wire to complete the circuit.

With the salt bridge in place, electrons can start to flow through the same basic redox reaction as the one I show you at the beginning of this section. Zinc is being oxidized, releasing electrons that flow through the wire to the copper electrode, where they're available for the Cu^{2+} ions to use in forming copper metal. Copper ions from the copper(II) sulfate solution are being plated out on the copper electrode, while the zinc electrode is being consumed. The cations in the salt bridge migrate to the container with the copper electrode to replace the copper ions being consumed, while the anions in the salt bridge migrate toward the zinc side, where they keep the solution containing the newly formed Zn^{2+} cations electrically neutral.

The zinc electrode is called the *anode,* the electrode at which oxidation takes place, and it's labeled with a – sign. The copper electrode is called the *cathode,* the electrode at which reduction takes place, and it's labeled with a + sign.

This Daniell cell produces a little over 1 volt. You can get just a little more voltage if you make the solutions that the electrodes are in very concentrated. But what can you do if you want, for example, 2 volts? You have a couple of choices: You can hook two of these cells up together and produce 2 volts, or you can choose two different metals that are farther apart than zinc and copper on the activity series chart (see Chapter 7). The farther apart the metals are on the activity series, the more voltage the cell produces.

Electrolytic cells: Getting chemical reactions from electricity

An *electrolytic cell* uses electricity to produce a desired redox reaction. For instance, water can be decomposed by the use of electricity in an electrolytic cell. The overall cell reaction is

$$2\ H_2O(l) \rightarrow 2\ H_2(g) + O_2(g)$$

In a similar fashion, you can produce sodium metal and chlorine gas by the electrolysis of molten sodium chloride.

Producing chemical changes by passing an electric current through an electrolytic cell is called *electrolysis*. This reaction may be the recharging of a battery (as you see the next section) or one of many other applications. For instance, ever wonder how the aluminum in that aluminum can is mined? Aluminum ore is primarily aluminum oxide (Al_2O_3). People produce aluminum metal by reducing the aluminum oxide in a high-temperature electrolytic cell using approximately 250,000 amps. That's a lot of electricity. Taking old aluminum cans, melting them down, and reforming them into new cans is far cheaper than extracting the metal from the ore. That's why the aluminum industry is strongly behind the recycling of aluminum. It's just good business.

Electrolytic cells are also used in a process called *electroplating*. In electroplating, a more-expensive metal is plated (deposited in a thin layer) onto the surface of a cheaper metal by electrolysis. Back before plastic auto bumpers became popular, chromium metal was electroplated onto steel bumpers. Those five-dollar gold chains you can buy are really made of some cheap metal with an electroplated surface of gold.

Having it both ways with rechargeable batteries

Redox reactions can be reversed to regenerate the original reactants, allowing people to make rechargeable batteries. Nickel-cadmium (Ni-Cad) and lithium batteries fall into this category, but the most familiar type of rechargeable battery is probably the automobile battery.

The ordinary automobile battery, or *lead storage battery*, consists of six cells connected in series. The anode of each cell (where oxidation takes place) is lead, and the cathode (where reduction takes place) is lead dioxide (PbO_2). The electrodes are immersed in a sulfuric acid (H_2SO_4) solution. When you start your car, the following cell reactions take place:

- ✓ **Anode:** $Pb(s) + H_2SO_4(aq) \rightarrow PbSO_4(s) + 2 H^+ + 2 e^-$

- ✓ **Cathode:** $2 e^- + 2 H^+ + PbO_2(s) + H_2SO_4(aq) \rightarrow PbSO_4(s) + 2 H_2O(l)$

- ✓ **Overall reaction:** $Pb(s) + PbO_2(s) + 2 H_2SO_4(aq) \rightarrow 2 PbSO_4 + 2 H_2O(l)$

When this happens, both electrodes become coated with solid lead(II) sulfate, and the sulfuric acid is used up.

After you start the automobile, the alternator or generator takes over the job of producing electricity (for spark plugs, lights, and so on) and also recharges the battery. During charging, the automobile battery acts like an electrolytic cell. The alternator reverses both the flow of electrons into the battery and the original redox reactions, and it regenerates the lead and lead dioxide:

$$2 PbSO_4(s) + 2 H_2O(l) \rightarrow Pb(s) + PbO_2(s) + 2 H_2SO_4(aq)$$

The lead storage battery can be discharged and charged many times. But the shock of running over bumps in the road or into the curb flakes off a little of the lead(II) sulfate and eventually causes the battery to fail.

Chapter 9

Measuring Substances with the Mole

* *

In This Chapter

▶ Figuring out how to count by weighing

▶ Understanding the mole concept

▶ Using the mole in chemical calculations

* *

*C*hemists make new substances in a process called *synthesis.* And a logical question they ask is "how much?" — how much of this reactant do I need to make this much product? How much product can I make with this much reactant?

To answer these questions, chemists must be able to take a balanced chemical equation, expressed in terms of atoms and molecules, and convert it to grams or pounds or tons — some type of unit they can actually weigh out in the lab. The mole concept enables chemists to move from the microscopic world of atoms and molecules to the real world of grams and kilograms, and it's one of the most important central concepts in chemistry. In this chapter, I introduce you to Mr. Mole.

Counting by Weighing

Counting by weighing is one of the most efficient ways of counting large numbers of objects. Suppose that you have a job packing 1,000 nuts and 1,000 bolts in big bags, and you get paid for each bag you fill. So what's the most efficient and quickest way of counting out nuts and bolts? Weigh out a hundred, or even ten, of each and then figure out how much a thousand of each will weigh. Fill up the bag with nuts until it weighs the amount you figured for 1,000 nuts. After you have the correct amount of nuts, use the same process to fill the bag with bolts.

In chemistry, you count very large numbers of particles, such as atoms and molecules. To count them efficiently and quickly, you use the count-by-weighing method, which means you need to know how much individual atoms and molecules weigh. Here's where you find the weights:

- ✔ **Atoms:** Get the weights of the individual atoms on the periodic table — just find the atomic mass number.

- ✔ **Compounds:** Simply add together the weights of the individual atoms in the compound to figure the molecular weight or formula weight. (*Note: Molecular weights* refer to covalently bonded compounds, and *formula weights* refer to both ionic and covalent compounds.)

Here's a simple example that shows how to calculate the molecular weight of a compound: Water, H_2O, is composed of two hydrogen atoms and one oxygen atom. By looking on the periodic table, you find that one hydrogen atom weights 1.0079 amu and one oxygen atom weighs 15.999 amu (*amu* stands for *atomic mass units* — see Chapter 2 for details). To calculate the molecular weight of water, simply add together the atomic weights of two hydrogen atoms and one oxygen atom:

Two hydrogen atoms: 2×1.0079 amu = 2.016 amu

One oxygen atom: 1×15.999 amu = 15.999 amu

Weight of the water molecule: 2.016 amu + 15.999 amu = 18.015 amu

Now try a little harder one. Calculate the formula weight of aluminum sulfate, $Al_2(SO_4)_3$. In this salt, you have two aluminum atoms, three sulfur atoms, and twelve oxygen atoms. After you find the individual weights of the atoms on the periodic table, you can calculate the formula weight like this:

[2 aluminum atoms + 3 sulfur atoms + 12 oxygen atoms] = weight of $Al_2(SO_4)_3$

$[(2 \times 26.982$ amu$) + (3 \times 32.066$ amu$) + (12 \times 15.999$ amu$)]$ = 342.15 amu

Moles: Putting Avogadro's Number to Good Use

When people deal with objects, they often think in terms of a convenient amount. For example, when a woman buys earrings, she normally buys a pair of them. When a man goes to the grocery store, he buys eggs by the dozen. Likewise, when chemists deal with atoms and molecules, they need a convenient unit that takes into consideration the very small size of atoms and molecules. They use a unit called the *mole*.

Defining the mole

The word *mole* stands for a number — approximately 6.022×10^{23}. It's commonly called *Avogadro's number,* named for Amedeo Avogadro, the scientist who laid the groundwork for the mole principle.

Now a mole — 6.022×10^{23} — is a really big number. When written in longhand notation, it's

602,200,000,000,000,000,000,000

And *that* is why I like scientific notation. (If you had a mole of marshmallows, it'd cover the United States to a depth of about 600 miles. A mole of rice grains would cover the land area of the world to a depth of about 75 meters.)

Avogadro's number stands for a certain number of *things.* Normally, those things are atoms and molecules. So the mole relates to the microscopic world of atoms and molecules. But how does it relate to the macroscopic world where you work?

The answer is that a *mole* (abbreviated as *mol*) is equal to the following:

- **For carbon:** A mole is the number of atoms in exactly 12 grams of C-12, a particular isotope of carbon. So if you have exactly 12 grams of ^{12}C, you have 6.022×10^{23} carbon atoms, which is also a mole of ^{12}C atoms.

- **For any other element:** A mole is the atomic weight using grams instead of atomic mass units.

✔ **For a compound:** For a compound, a mole is the formula (or molecular) weight in grams instead of atomic mass units.

Calculating weight, particles, and moles

The mole is the bridge between the microscopic and macroscopic world:

6.022×10^{23} particles ↔ 1 mole ↔ atomic/formula weight in grams

If you have any one of the three things — particles, moles, or grams — then you can calculate the other two.

For example, the weight of a water molecule is 18.015 amu. Because a mole is the formula (or molecular) weight in grams of a compound, you can now say that the weight of a mole of water is 18.015 grams. You can also say that 18.015 grams of water contains 6.022×10^{23} H_2O molecules, or a mole of water. And the mole of water is composed of two moles of hydrogen and one mole of oxygen.

Suppose you want to know how many water molecules are in 5.50 moles of water. You can set up a problem like this:

5.50 mol $\times 6.022 \times 10^{23}$ molecules/mol $= 3.31 \times 10^{24}$ molecules

Or suppose that you want to know how many moles are in 25.0 grams of water. You can set up the problem like this:

$$\frac{25.0g\ H_2O}{1} \bullet \frac{1\ mol\ H_2O}{18.015g\ H_2O} = 1.39\ moles\ H_2O$$

You can even go from grams to particles by going through the mole. For example, how many molecules are in 100.0 grams of carbon dioxide? The first thing you have to do is determine the molecular weight of CO_2. Look at the periodic table to find that one carbon atom equals 12.011 amu and one oxygen atom weighs 15.000 amu. Now figure the molecular weight, like this:

$[(1 \times 12.011 \text{ g/mol}) + (2 \times 15.999 \text{ g/mol})] =$
$44.01 \text{ g/mol for } CO_2$

Now you can work the problem:

$$\frac{100.0\text{g } CO_2}{1} \cdot \frac{1 \text{ mol } CO_2}{44.01\text{g}} \cdot \frac{6.022 \cdot 10^{23} \text{ molecules}}{1 \text{ mol}} =$$
$$1.368 \cdot 10^{24} \ CO_2 \text{ molecules}$$

And going from particles to moles to grams is just as easy.

Finding formulas of compounds

You can use the mole concept to calculate the empirical formula of a compound using the *percentage composition* data for that compound — the percentage by weight of each element in the compound. (The *empirical formula* indicates the different types of elements in a molecule and the lowest whole-number ratio of each kind of atom in the molecule. See Chapter 6 for details.)

When I try to determine the empirical formula of a compound, I often have percentage data available. The determination of the percentage composition is one of the first analyses that a chemist does in learning about a new compound. Here's how to find an empirical formula using moles and the percentages of each element:

1. **Assume you have 100 grams of the compound so you can use the percentages as weights; then convert the weight of each element to moles.**

   For example, suppose you determine that a particular compound has the following weight percentage of elements present: 26.4 percent Na, 36.8 percent S, and 36.8 percent O. Because you're dealing with percentage data (amount per hundred), assume that you have 100 grams of the compound so you can write the percentages as weights. Then convert each mass to moles, like this:

$$\frac{26.4g\ Na}{1} \cdot \frac{1\ mol\ Na}{22.9g} = 1.15\ mol\ Na$$

$$\frac{36.8g\ S}{1} \cdot \frac{1\ mol\ S}{32.07g} = 1.15\ mol\ S$$

$$\frac{36.8g\ O}{1} \cdot \frac{1\ mol\ O}{16.0g} = 2.30\ mol\ O$$

2. **Write the empirical formula, changing subscripts to whole numbers if necessary.**

   Now you can write an empirical formula of $Na_{1.15}S_{1.15}O_{2.30}$. Your subscripts have to be whole numbers, so divide each of these by the smallest, 1.15, to get $NaSO_2$. (If a subscript is 1, it's not shown.)

You can then calculate a weight for the empirical formula by adding together the atomic masses on the periodic table of one sodium (Na), one sulfur (S) and two oxygen (O). This gives you an empirical formula weight of 87.056 grams.

Chemical Reactions and Moles

When you're working with chemical reactions, moles can help you figure out how much of a product you can expect to get based on how much of the reactants you have.

Take a look at my favorite reaction, the Haber process, which is a method of preparing ammonia (NH_3) by reacting nitrogen gas with hydrogen gas:

$$N_2(g) + 3\ H_2(g) \leftrightarrow 2\ NH_3(g)$$

In Chapter 7, I use this reaction over and over again for various examples (like I said, it's my *favorite* reaction) and explain that you can read the reaction like this: one molecule of nitrogen gas reacts with three molecules of hydrogen gas to yield two molecules of ammonia:

$$N_2(g) + 3\ H_2(g) \leftrightarrow 2\ NH_3(g)$$

1 molecule + 3 molecules $\leftrightarrow$ 2 molecules

If you like, you can scale everything up by a factor of 12:

$N_2(g) + 3 H_2(g) \leftrightarrow 2 NH_3(g)$

12 molecules + 3(12 molecules) ↔ 2(12 molecules)

1 dozen molecules + 3 dozen molecules ↔ 2 dozen molecules

Or how about a factor of 6.022×10^{23}? But wait a minute! Isn't 6.022×10^{23} a mole? So you can write the equation like this:

$N_2(g) + 3 H_2(g) \leftrightarrow 2 NH_3(g)$

6.022×10^{23} molecules + $3(6.022 \times 10^{23})$ molecules ↔ $2(6.022 \times 10^{23}$ molecules)

1 mole + 3 moles ↔ 2 moles

That's right — not only can those coefficients in the balanced chemical equation represent atoms and molecules, but they can also represent the number of moles.

If you know the formula weight of the reactants and product, you can calculate how much you need and how much you'll get. All you need to do is figure the molecular weights of each reactant and product and then incorporate the weights into the equation. Use the periodic table to find the weights of the atoms and the compound (see the earlier section "Counting by Weighing" for directions) and multiply those numbers by the number of moles, like this:

1 mol(28.014 g/mol) + 3 mol(2.016 g/mol) = 2 mol(17.031g/mol)

28.014 g of N_2 + 6.048 g of H_2 = 34.062 g of NH_3

Reaction stoichiometry

When you understand the weight relationships in a chemical reaction, you can do some stoichiometry problems. *Stoichiometry* refers to the mass relationship in chemical equations.

When you get ready to work stoichiometry types of problems, you must start with a balanced chemical equation. If you don't have it to start with, go ahead and balance the equation.

Look at my favorite reaction — you guessed it — the Haber process:

$$N_2(g) + 3 H_2(g) \leftrightarrow 2 NH_3(g)$$

Suppose that you want to know how many grams of ammonia can be produced from the reaction of 75.00 grams of nitrogen with excess hydrogen. The mole concept is the key. The coefficients in the balanced equation are not only the number of individual atoms or molecules but also the number of moles:

$$N_2(g) + 3 H_2(g) \leftrightarrow 2 NH_3(g)$$

1 mole + 3 moles $\leftrightarrow$ 2 moles

1 mol(28.014 g/mol) + 3 mol(2.016 g/mol) = 2 mol(17.031g/mol)

First, convert the 75.00 grams of nitrogen to moles of nitrogen. Then use the ratio of the moles of ammonia to the moles of nitrogen from the balanced equation to convert to moles of ammonia. Finally, take the moles of ammonia and convert that number to grams. The equation looks like this:

$$\frac{75.00g\ N_2}{1} \cdot \frac{1\ mol\ N_2}{28.014g\ N_2} \cdot \frac{2\ mol\ NH_3}{1\ mol\ N_2} \cdot \frac{17.031g\ NH_3}{1\ mol\ NH_3}$$

$$= 91.19g\ NH_3$$

A *stoichiometric ratio* — such as mol NH_3/mol N_2 — enables you to convert from the moles of one substance in a balanced chemical equation to the moles of another substance.

Percent yield

In almost any reaction, you're going to produce less of the product than you expected. You may produce less because most reactions are equilibrium reactions (see Chapter 8), because of sloppy technique or impure reactants, or because some other conditions come into play. Chemists can get an idea of the efficiency of a reaction by calculating the *percent yield* for the reaction using this equation:

$$\%\ yield = \frac{actual\ yield}{theoretical\ yield} \cdot 100\%$$

The *actual yield* is how much of the product you get when you carry out the reaction. The *theoretical yield* is how much of the product you calculate you'll get. The ratio of these two yields gives you an idea about how efficient the reaction is.

For the reaction of rust to iron (see the preceding section), your theoretical yield is 699.5 grams of iron; suppose your actual yield is 525.0 grams. Therefore, the percent yield is

$$\% \text{ yield} = \frac{525.0g}{699.5g} \bullet 100\% = 75.05\%$$

A percent yield of about 75 percent isn't too bad, but chemists and chemical engineers would rather see 90+ percent. One industrial plant using the Haber reaction has a percent yield of better than 99 percent. Now that's efficiency!

Limiting reactants

In a chemical reaction, you normally run out of one of the reactants and have some others left over. The reactant you run out of first is called the *limiting reactant,* and it determines the amount of product formed. (In some of the problems sprinkled throughout this chapter, I tell you which reactant is the limiting one by saying you have *an excess* of the other reactant(s).)

In this section, I show you how to calculate which reactant is the limiting reactant.

Here is a reaction between ammonia and oxygen.

$$4\,NH_3(g) + 5\,O_2(g) \rightarrow 4\,NO(g) + 6\,H_2O(g)$$

Suppose that you start out with 100.0 grams each of both ammonia and oxygen, and you want to know how many grams of NO (nitrogen monoxide, sometimes called *nitric oxide*) you can produce. You must determine the limiting reactant and base your stoichiometric calculations on it.

To figure out which reactant is the limiting reactant, you calculate the mole-to-coefficient ratio: Calculate the number of moles of both ammonia and oxygen, and then divide each by their coefficients in the balanced chemical equation. The one with the smallest mole-to-coefficient ratio is the limiting reactant. For the reaction of ammonia to nitric oxide, you can calculate the mole-to-coefficient ratio for the ammonia and oxygen like this:

$$\frac{100.0g\ NH_3}{1} \bullet \frac{1\ mol\ NH_3}{17.03g} = 5.87\ mol \div 4 = 1.47\ for\ the\ NH_3$$

$$\frac{100.0g\ O_2}{1} \bullet \frac{1\ mol\ O_2}{32.00g} = 3.13\ mol \div 5 = 0.625\ for\ the\ O_2$$

Ammonia has a mole-to-coefficient ratio of 1.47, and oxygen has a ratio of 0.625. Because oxygen has the lowest ratio, oxygen is the limiting reactant, and you need to base your calculations on it.

$$\frac{100.0g\ O_2}{1} \bullet \frac{1\ mol\ O_2}{32.00g} \bullet \frac{4\ mol\ NO}{5\ mol\ O_2} \bullet \frac{30.01g\ NO}{1\ mol\ NO} = 75.02g\ NO$$

That 75.02 grams NO is your theoretical yield. But you can even calculate the amount of ammonia left over. You can figure the amount of ammonia consumed with this equation:

$$\frac{100.0g\ O_2}{1} \bullet \frac{1\ mol\ O_2}{32.00g} \bullet \frac{4\ mol\ NO}{5\ mol\ O_2} \bullet \frac{17.03g\ NH_3}{1\ mol\ NH_3} = 42.58g\ NH_3$$

You started with 100.0 grams of ammonia, and you used 42.58 grams of it. The difference (100 grams - 42.58 grams = 57.42 grams) is the amount of ammonia left over.

Chapter 10

A Salute to Solutions

*I*n this chapter, I show you some of the properties of solutions. I introduce you to the different ways chemists represent a solution's concentration, and I tell you about the colligative properties of solutions (properties that depend on the number of solute particles) and relate them to ice cream–making and antifreeze. I also introduce you to colloids, close cousins of solutions, and discuss the importance of particle size. So sit back, sip on your solution of choice, and read all about solutions.

Mixing Things Up with Solutes, Solvents, and Solutions

A *solution* is a homogeneous mixture, meaning that it's the same throughout. If you dissolve sugar in water and mix it really well, for example, your mixture is basically the same no matter where you sample it.

A solution is composed of a solvent and one or more solutes. The *solvent* is the substance that's present in the largest amount, and the *solute* is the substance that's present in the lesser amount. These definitions work most of the time, but there are a few cases of extremely soluble salts, such as lithium chloride, in which more than 5 grams of salt can be dissolved in 5 milliliters of water. However, water is still considered the solvent, because it's the species that hasn't changed state.

In addition, there can be more than one solute in a solution. You can dissolve salt in water to make a brine solution, and then you can dissolve some sugar in the same solution. You then have two solutes, salt and sugar, but you still have only one solvent — water.

Most people think of liquids when they think about solutions, but you can also have solutions of gases or solids. Earth's atmosphere, for example, is a solution. Because air is almost 79 percent nitrogen, it's considered the solvent, and the oxygen, carbon dioxide, and other gases are considered the solutes. *Alloys* are solutions of one metal in another metal. Brass is a solution of zinc in copper.

How dissolving happens

Why do some things dissolve in one solvent and not another? For example, oil and water don't mix to form a solution, but oil does dissolve in gasoline.

There's a general rule of solubility that says *like-dissolves-like* in regards to polarity of both the solvent and solutes. A *polar* material is composed of covalent bonds with a positive and negative end of the molecule. (For a rousing discussion of water and its polar covalent bonds, see Chapter 6.) A polar substance such as water dissolves polar solutes, such as salts and alcohols. Oil, however, is composed of largely nonpolar bonds, so water doesn't act as a suitable solvent for oil.

Concentration limits

Solubility is the maximum amount of solute that will dissolve in a given amount of a solvent at a specified temperature. You know from your own experiences, I'm sure, that there's a limit to how much solute you can dissolve in a given amount of solvent. Most people have been guilty of putting far too much sugar in iced tea. No matter how much you stir, some undissolved sugar stays at the bottom of the glass.

Solubility normally has the units of grams solute per 100 milliliters of solvent (g/100 mL), and it's related to the temperature of the solvent:

- ✔ **Solids in liquids:** For solids dissolving in liquids, solubility normally increases with increasing temperature. For instance, if you heat that iced tea, the sugar at the bottom will readily dissolve.

- ✔ **Gases in liquids:** For gases dissolving in liquids, such as oxygen dissolving in lake water, the solubility goes down as the temperature increases. This is the basis of *thermal pollution,* the addition of heat to water that decreases the solubility of the oxygen and affects the aquatic life.

Saturated facts

A *saturated* solution contains the maximum amount of dissolved solute possible at a given temperature. If it has less than this amount, it's called an *unsaturated* solution. Sometimes, under unusual circumstances, the solvent may actually dissolve more than its maximum amount and become *supersaturated.* This supersaturated solution is unstable, though, and sooner or later, the solute will *precipitate* (form a solid) until the saturation point has been reached.

If a solution is unsaturated, then the amount of solute that's dissolved may vary over a wide range. A couple of rather nebulous terms describe the relative amount of solute and solvent that you can use:

- ✔ **Dilute:** You can say that the solution is *dilute,* meaning that relatively speaking, there's very little solute per given amount of solvent. If you dissolve 0.01 grams of sodium chloride in a liter of water, for example, the solution is dilute. (Another dilute solution? A $1 margarita, as one of my students once pointed out — that's a lot of solvent [water] and very little solute [tequila].)

- ✔ **Concentrated:** A *concentrated* solution contains a large amount of solute per the given amount of solvent. If you dissolve 200 grams of sodium chloride in a liter of water, for example, the solution is concentrated.

But suppose you dissolve 25 grams or 50 grams of sodium chloride in a liter of water. Is the solution dilute or concentrated? These terms don't hold up very well for most cases.

Understanding Solution Concentration Units

People need to have a quantitative method to describe the relative amount of solute and solvent in a solution. For instance, consider the case of IV solutions — they must have a very precise amount of solute in them, or the patient will be in danger. Chemists use *solution concentration units* to quantify solutes and solvents.

You can use a variety of solution concentration units to quantitatively describe the relative amounts of the solute(s) and the solvent. In everyday life, percentage is commonly used. In chemistry, *molarity* (the moles of solute per liter of solution) is the solution concentration unit of choice. In certain circumstances, though, another unit, *molality* (the moles of solute per kilogram of solvent), is used. And I use parts-per-million or parts-per-billion when I discuss pollution control. The following sections cover these concentration units.

Percent composition

You've probably seen labels like "5% acetic acid" on a bottle of vinegar, "3% hydrogen peroxide" on a bottle of hydrogen peroxide, or "5% sodium hypochlorite" on a bottle of bleach. Those percentages are expressing the concentration of that particular solute in each solution. *Percentage* is the amount per 100. Depending on the way you choose to express the percentage, the units of amount per one hundred vary. Three different percentages are commonly used:

 ✔ Weight/weight (w/w) percentage

 ✔ Weight/volume (w/v) percentage

 ✔ Volume/volume (v/v) percentage

Unfortunately, although the percentage of solute is often listed, the method (w/w, w/v, v/v) is not. In this case, I normally assume that the method is weight/weight, but I'm sure you know about assumptions.

Most of the solutions I talk about in the following examples of these percentages are *aqueous* solutions, solutions in which water is the solvent.

Weight/weight percentage

In *weight/weight percentage*, or *weight percentage*, the weight of the solute is divided by the weight of the solution and then multiplied by 100 to get the percentage. Normally, the weight unit is grams. Mathematically, it looks like this:

$$w/w\% = \frac{\text{grams solute}}{\text{grams solution}} \bullet 100\%$$

If, for example, you dissolve 5.0 grams of sodium chloride in 45 grams of water, the weight percent is

$$w/w\% = \frac{5.0\text{g NaCl}}{50 \text{ grams solution}} \bullet 100\% = 10.\%$$

Therefore, the solution is a 10 percent (w/w) solution.

Weight percentage is the easiest percentage solution to make. Suppose that you want to make 350.0 grams of a 5 percent (w/w) sucrose (table sugar) solution. You know that 5 percent of the weight of the solution is sugar, so you can multiply the 350.0 grams by 0.05 to get the weight of the sugar:

350.0 grams × 0.05 = 17.5 grams of sugar

The rest of the solution (350.0 grams − 17.5 grams = 332.5 grams) is water. You can simply weigh out 17.5 grams of sugar and add it to 332.5 grams of water to get your 5 percent (w/w) solution.

Weight/volume percentage

Weight/volume percentage is very similar to weight/weight percentage, but instead of using grams of solution in the denominator, it uses milliliters of solution:

$$w/v\% = \frac{\text{grams solute}}{\text{mL solution}} \bullet 100\%$$

Suppose that you want to make 100 milliliters of a 15 percent (w/v) potassium nitrate solution. Because you're making 100 milliliters, you already know that you're going to weigh out 15 grams of potassium nitrate (commonly called saltpeter — KNO_3).

Now, here comes something that's a little different: You're more concerned with the final volume of the solution than the amount of solvent you use. So you dissolve the 15 grams of KNO_3 in a little bit of water and dilute it to exactly 100 milliliters in a volumetric flask. In other words, you dissolve and dilute 15 grams of KNO_3 to 100 milliliters. (I tend to abbreviate *dissolve and dilute* by writing *d & d*.) You won't know exactly how much water you put in, but it's not important as long as the final volume is 100 milliliters.

You can also use the percentage and volume to calculate the grams of solute present. You may want to know how many grams of sodium hypochlorite are in 500 milliliters of a 5 percent (w/v) solution of household bleach. You can set up the problem like this:

$$\frac{5g\ NaOCl}{100mL\ solution} \bullet \frac{500mL\ solution}{1} = 25g\ NaOCl$$

You now know that you have 25 grams of sodium hypochlorite in the 500 milliliters of solution.

Volume/volume percentage

If both the solute and solvent are liquids, using a volume/volume percentage is convenient. With *volume/volume percentages*, both the solute and solution are expressed in milliliters:

$$v/v\% = \frac{mL\ solute}{mL\ solution} \bullet 100\%$$

Ethyl alcohol (the drinking alcohol) solutions are commonly made using volume/volume percentages. If you want to make 100 milliliters of a 50 percent ethyl alcohol solution, you take 50 milliliters of ethyl alcohol and dilute it to 100 milliliters with water. Again, it's a case of dissolving and diluting to the required volume.

You can't simply add 50 milliliters of alcohol to 50 milliliters of water — you'd get less than 100 milliliters of solution. The polar water molecules attract the polar alcohol molecules. This tends to fill in the open framework of water molecules and prevents the volumes from simply being added together.

Molarity: Comparing solute to solution

Molarity is the concentration unit chemists use most often, because it utilizes moles. The mole concept is central to chemistry, and molarity lets chemists easily work solutions into reaction stoichiometry. (If you're wondering what burrowing, insect-eating mammals have to do with chemistry, let alone what stoichiometry is, just flip to Chapter 9 for the scoop.)

Molarity (M) is defined as the moles of solute per liter of solution. Mathematically, it looks like this:

$$M = \frac{\text{mol solute}}{\text{L solution}}$$

For example, you can take 1 mole (abbreviated as *mol*) of KCl (formula weight of 74.55 g/mol; see Chapter 10) and dissolve and dilute the 74.55 grams to 1 liter of solution in a volumetric flask. You then have a 1-molar solution of KCl. You can label that solution as 1 M KCl.

When preparing molar solutions, always dissolve and dilute to the required volume. So to dissolve 74.55 grams of KCl to 1 liter of solution, you don't add the 74.55 grams to 1 liter of water. You want to end up with a final volume of 1 liter.

Here's another example: If 25.0 grams of KCl are dissolved and diluted to 350.0 milliliters, how would you calculate the molarity of the solution? You know that molarity is moles of solute per liter of solution. So you can take the grams, convert them to moles using the formula weight of KCl (74.55 g/mol), and divide them by 0.350 liters (350.0 milliliters). You can set up the equation like this:

$$\frac{25.0\text{g KCl}}{1} \cdot \frac{1\text{ mol KCl}}{74.55\text{g}} \cdot \frac{1}{0.350\text{ L}} = 0.958\text{ M}$$

Now suppose that you want to prepare 2.00 liters of a 0.550 M KCl solution. The first thing you do is calculate how much KCl you need to weigh:

$$\frac{0.550 \text{ mol KCl}}{L} \bullet \frac{74.55 \text{g KCl}}{1 \text{ mol}} \bullet \frac{2.00 \text{ L}}{1} = 82.0 \text{g KCl}$$

You then take that 82.0 grams of KCl and dissolve and dilute it to 2.00 liters.

Diluting solutions to the right molarity

There's one more way to prepare solutions — the dilution of a more concentrated solution to a less-concentrated one. For example, you can buy hydrochloric acid from the manufacturer as a concentrated solution of 12.0 M. Suppose that you want to prepare 500 milliliters of 2.0 M HCl. You can dilute some of the 12.0 M to 2.0 M, but how much of the 12.0 M HCl do you need? You can easily figure the volume (V) you need by using the following formula:

$$V_{old} \times M_{old} = V_{new} \times M_{new}$$

In the preceding equation, V_{old} is the old volume, or the volume of the original solution, M_{old} is the molarity of the original solution, V_{new} is the volume of the new solution, and M_{new} is the molarity of the new solution. After substituting the values, you have the following:

$$V_{old} \times 12.0 \text{ M} = 500.0 \text{ milliliters} \times 2.0 \text{ M}$$

$$V_{old} = (500.0 \text{ milliliters} \times 2.0 \text{ M})/12.0 \text{ M} = 83.3 \text{ milliliters}$$

You then take 83.3 milliliters of the 12.0 M HCl solution and dilute it to exactly 500.0 milliliters.

If you're actually doing a dilution of concentrated acids, be sure to *add the acid to the water* instead of the other way around! If the water is added to the concentrated acid, then so much heat will be generated that the solution will quite likely splatter all over you. So to be safe, you should take about 400 milliliters of water, slowly add the 83.3 milliliters of the concentrated HCl as you stir, and then dilute to the final 500 milliliters with water.

Molarity in stoichiometry: Figuring out how much you need

The usefulness of the molarity concentration unit is readily apparent when dealing with reaction stoichiometry. For example, suppose that you want to know how many milliliters of 2.50 M sulfuric acid it takes to neutralize a solution containing 100.0 grams of sodium hydroxide. The first thing you must do is write the balanced chemical equation for the reaction:

$$H_2SO_4(aq) + 2\,NaOH(aq) \rightarrow 2\,H_2O(l) + Na_2SO_4(aq)$$

You know that you have to neutralize 100.0 grams of NaOH. You can convert the weight to moles (using the formula weight of NaOH, 40.00 g/mol) and then convert from moles of NaOH to moles of H_2SO_4. Then you can use the molarity of the acid solution to get the volume:

$$\frac{100.0g\ NaOH}{1} \cdot \frac{1\ mol\ NaOH}{40.00g} \cdot \frac{1\ mol\ H_2SO_4}{2\ mol\ NaOH} \cdot \frac{L}{2.5\ mol\ H_2SO_4}$$

$$\cdot \frac{1000\ mL}{1\ L} = 500.0\ mL$$

It takes 500.0 milliliters of the 2.50 M H_2SO_4 solution to completely react with the solution that contains 100 grams of NaOH.

Molality: Comparing solute to solvent

Molality is another concentration term that involves moles of solute. It isn't used very much, but you may happen to run across it. *Molality (m)* is defined as the moles of solute per kilogram of solvent. It's one of the few concentration units that doesn't use the total solution's weight or volume. Mathematically, it looks like this:

$$m = \frac{mol\ solute}{Kg\ solvent}$$

Suppose, for example, you want to dissolve 15.0 grams of NaCl in 50.0 grams of water. You can calculate the molality like this (you must convert the 50.0 grams to kilograms before you use it in the equation):

$$\frac{15.0 \text{g NaCl}}{1} \bullet \frac{1 \text{ mol}}{58.44 \text{g NaCl}} \bullet \frac{1}{0.0500 \text{ Kg}} = 5.13 \text{m}$$

Parts per million

Percentage and molarity, and even molality, are convenient units for the solutions that chemists routinely make in the lab or the solutions commonly found in nature. However, if you begin to examine the concentrations of certain pollutants in the environment, you find that those concentrations are very, very small. Percentage and molarity work when you're measuring solutions in the environment, but they're not very convenient. To express the concentrations of very dilute solutions, scientists have developed another concentration unit — parts per million.

Percentage is parts per hundred, or grams solute per 100 grams of solution. *Parts per million (ppm)* is grams of solute per 1 million grams of solution — or as it's most commonly expressed, milligrams of solute per kilogram of solution, which is the same ratio. It's expressed this way because chemists can easily weigh out milligrams or even tenths of milligrams, and if you're talking about aqueous solutions, a kilogram of solution is the same as a liter of solution. (The density of water is 1 gram per milliliter, or 1 kilogram per liter. The weight of the solute in these solutions is so very small that it's negligible when converting from the mass of the solution to the volume.)

By law, the maximum contamination level of lead in drinking water is 0.05 ppm. This number corresponds to 0.05 milligrams of lead per liter of water. That's pretty dilute. But mercury is regulated at the 0.002 ppm level. Sometimes, even this unit isn't sensitive enough, so environmentalists have resorted to the parts per billion (ppb) or parts per trillion (ppt) concentration units. Some neurotoxins are deadly at the parts per billion level.

Chapter 11

Acids and Bases

• •

In This Chapter

▶ Discovering the properties of acids and bases

▶ Finding out about the acid-base theory

▶ Differentiating between strong and weak acids and bases

▶ Understanding indicators

▶ Taking a look at the pH scale

• •

*W*alk into any kitchen or bathroom, and you find a
multitude of acids and bases. Open the refrigerator,
and you find soft drinks full of carbonic acid. In the pantry,
you find vinegar and baking soda, an acid and a base. Peek
under the sink, and you notice the ammonia and other cleaners, most of which are bases. Check out that can of lye-based
drain opener — it's highly basic. In the medicine cabinet, you
find aspirin, an acid, and antacids of all types. Your everyday
world is full of acids and bases — and so is the everyday
world of the industrial chemist. In this chapter, I cover acids
and bases, indicators and pH, and some good *basic* chemistry.

Observing Properties of Acids and Bases

Table 11-1 lists the properties of acids and bases that you can
observe in the world around you.

Table 11-1	Properties of Acids and Bases	
Property	*Acid*	*Base*
Taste (but remember: in the lab, you test, not taste!)	Sour	Bitter
Feel on the skin	Produces a painful sensation	Feels slippery
Reactions	Reacts with certain metals (magnesium, zinc, and iron) to produce hydrogen gas Reacts with limestone and baking soda to produce carbon dioxide	Reacts with oils and greases React with acids to produce a salt and water
Reaction with litmus paper	Turns the paper red	Turns the paper blue

The Brønsted-Lowry Acid-Base Theory

Here's how the Brønsted-Lowry theory defines acids and bases:

- ✔ An *acid* is a proton (H⁺) donor
- ✔ A *base* is a proton (H⁺) acceptor

The base accepts the H⁺ by furnishing a lone pair of electrons for a *coordinate-covalent* bond, which is a covalent bond (shared pair of electrons) in which one atom furnishes both of the electrons for the bond. Normally, one atom furnishes one electron for the bond and the other atom furnishes the second electron (see Chapter 6). In the coordinate-covalent bond, one atom furnishes both bonding electrons.

Figure 11-1 shows the NH_3/HCl reaction using the electron-dot structures of the reactants and products. (I cover electron-dot structures in Chapter 7.)

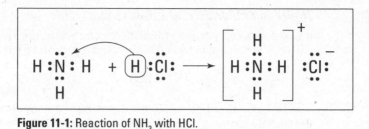

Figure 11-1: Reaction of NH_3 with HCl.

HCl is the proton donor, and the acid and ammonia are the proton acceptor, or the base. Ammonia has a lone pair of nonbonding electrons that it can furnish for the coordinate-covalent bond.

Understanding Strong and Weak Acids and Bases

I want to introduce you to a couple of different categories of acids and bases — strong and weak. *Strength* refers to the amount of ionization or breaking apart that a particular acid or base undergoes.

Acid-base strength is not the same as concentration. *Concentration* refers to the amount of acid or base that you initially have. You can have a concentrated solution of a weak acid, or a dilute solution of a strong acid, or a concentrated solution of a strong acid or . . . well, I'm sure you get the idea.

Remember, as I said previously, an acid yields hydronium ions (H_3O^+) and a base reacts with them. Many times that base will be the hydroxide ion (OH^-). The concentrations of these acid and base ions will prove to be important to us.

Strong: Ionizing all the way

An acid or base is *strong* if it ionizes completely — the reactants keep creating the product until they're all used up. In this section, I discuss both strong acids and strong bases.

Hydrogen chloride and other strong acids

If you dissolve hydrogen chloride gas in water, the HCl reacts with the water molecules and donates a proton to them:

$$HCl(g) + H_2O(l) \rightarrow Cl^- + H_3O^+$$

The H_3O^+ ion is called the hydronium ion. This reaction goes essentially to completion. In this case, all the HCl ionizes to H_3O^+ and Cl^- until there's no more HCl present. Because HCl ionizes essentially 100 percent in water, it's considered a strong acid. Note that water, in this case, acts as a base, accepting the proton from the hydrogen chloride.

Calculating ion concentration in solutions

Because strong acids ionize completely, calculating the concentration of the ions in solution is easy if you know the initial concentration of the strong acid. For example, suppose that you bubble 0.1 moles of HCl gas into a liter of water (see Chapter 9 to get a firm grip on moles). You can say that the initial concentration of HCl is 0.1 M (0.1 mol/L). *M* stands for molarity, and *mol/L* stands for moles of solute per liter. (For a detailed discussion of molarity and other concentration units, see Chapter 10.)

You can represent this 0.1 M concentration for the HCl in this fashion: [HCl] = 0.1. Here, the brackets around the compound indicate molar concentration, or mol/L. Because the HCl completely ionizes, you see from the balanced equation that for every HCl that ionizes, you get one hydronium ion and one chloride ion. So the concentration of ions in that 0.1 M HCl solution is

$$[H_3O^+] = 0.1, \text{ and } [Cl^-] = 0.1$$

The relationship between the amount of acid you start with and the amount of acid or base ions you end up with is valuable when you calculate the pH of a solution. (And you can do just that in the section "Phun with the pH Scale," later in this chapter.)

Looking at common strong acids

Table 11-2 lists the most common strong acids you're likely to encounter. Most of the other acids you encounter are weak.

Table 11-2	Common Strong Acids
Name	*Formula*
Hydrochloric acid	HCl
Hydrobromic acid	HBr
Hydroiodic acid	HI
Nitric acid	HNO_3
Perchloric acid	$HClO_4$
Sulfuric acid (first ionization only)	H_2SO_4

Sulfuric acid is called a *diprotic* acid. It can donate two protons, but only the first ionization goes 100 percent. The other acids listed in Table 11-2 are *monoprotic* acids, because they donate only one proton.

Strong bases: Hydroxide ions

Strong bases are those compounds that totally dissociate in water, yielding some cation and the hydroxide ion. It is the hydroxide ion that we normally refer to as the base, because it is what accepts the proton. Calculating the hydroxide ion concentration is really straightforward. Suppose that you have a 1.5 M (1.5 mol/L) NaOH solution. The sodium hydroxide, a salt, completely *dissociates* (breaks apart) into ions:

$$NaOH \rightarrow Na^+ + OH^-$$

If you start with 1.5 mol/L NaOH, then you have the same concentration of ions:

$$[Na^+] = 1.5 \text{ and } [OH^-] = 1.5$$

Weak: Ionizing partially

A *weak* acid or base ionizes only partially. The reactants aren't completely used up creating the products, as they are with strong acids and bases. Instead, the reactants establish equilibrium. In *equilibrium systems,* two exactly opposite chemical reactions — one on each side of the reaction arrow — are occurring at the same place, at the same time, with the same speed of reaction. (For a discussion of equilibrium systems, see Chapter 7.)

Acetic acid and other weak acids

Suppose that you dissolve acetic acid (CH_3COOH) in water. It reacts with the water molecules, donating a proton and forming hydronium ions. It also establishes equilibrium, where you have a significant amount of unionized acetic acid.

The acetic acid reaction with water looks like this:

$$CH_3COOH(l) + H_2O(l) \leftrightarrow CH_3COO^- + H_3O^+$$

The amount of hydronium ion that you get in solutions of acids that don't ionize completely is much less than it is with a strong acid. Acids that only partially ionize are called *weak acids*. In the case of acetic acid, about 5 percent ionizes, and 95 percent remains in the molecular form.

Calculating the hydronium ion concentration in weak acid solutions isn't as straightforward as it is in strong solutions, because not all the weak acid that dissolves initially has ionized. To calculate the hydronium ion concentration, you must use the equilibrium constant expression for the weak acid. Chapter 7 covers the K_{eq} expression that represents the equilibrium system. For weak acid solutions, you use a modified equilibrium constant expression called the K_a — the *acid ionization constant*.

One way to distinguish between strong and weak acids is to look for an acid ionization constant (K_a) value. If the acid has a K_a value, then it's weak.

Take a look at the generalized ionization of some weak acid HA:

$$HA + H_2O \leftrightarrow A^- + H_3O^+$$

The K_a expression for this weak acid is

$$K_a = \frac{[H_3O^+][A^-]}{[HA]}$$

Note that the *[HA]* represents the molar concentration of HA *at equilibrium,* not initially. Also, note that the concentration of water doesn't appear in the K_a expression, because there's so much that it actually becomes a constant incorporated into the K_a expression.

Now go back to that acetic acid equilibrium. The K_a for acetic acid is 1.8×10^{-5}. The K_a expression for the acetic acid ionization is

$$K_a = 1.8 \bullet 10^{-5} = \frac{\left[H_3O^+ \right]\left[CH_3COO^- \right]}{\left[CH_3COOH \right]}$$

You can use this K_a when calculating the hydronium ion concentration in, say, a 2.0 M solution of acetic acid. You know that the initial concentration of acetic acid is 2.0 M. You know that a little bit has ionized, forming a little hydronium ion and acetate ion. You also can see from the balanced reaction that for every hydronium ion that's formed, an acetate ion is also formed — so their concentrations are the same. You can represent the amount of $[H_3O^+]$ and $[CH_3COO^-]$ as x, so

$[H_3O^+] = [CH_3COO^-] = x$

To produce the x amount of hydronium and acetate ion, the same amount of ionizing acetic acid is required. So you can represent the amount of acetic acid remaining at equilibrium as the amount you started with, 2.0 M, minus the amount that ionizes, x:

$[CH_3COOH] = 2.0 - x$

For the vast majority of situations, you can say that x is very small in comparison to the initial concentration of the weak acid. So you can say that $2.0 - x$ is approximately equal to 2.0. This means that you can often approximate the equilibrium concentration of the weak acid with its initial concentration. The equilibrium constant expression now looks like this:

$$K_a = 1.8 \bullet 10^{-5} = \frac{[x][x]}{[2.0]} = \frac{[x]^2}{[2.0]}$$

At this point, you can solve for x, which is the $[H_3O^+]$:

$(1.8 \times 10^{-5})[2.0] = [x]^2$

$6.0 \times 10^{-3} = [H_3O^+]$

Weak bases: Ammonia

Weak bases, like weak acids, react with water to establish an equilibrium system. Ammonia is a typical weak base. It reacts with water to form the ammonium ion and the hydroxide ion:

$$NH_3(g) + H_2O(l) \rightleftarrows NH_4^+ + OH^-$$

Like a weak acid, a weak base is only partially ionized. There's a modified equilibrium constant expression for weak bases — the K_b. You use it exactly the same way you use the K_a (see "Acetic acid and other weak acids" for the details), except you solve for the [OH⁻].

Acid-Base Reactions: Using the Brønsted-Lowry System

The Brønsted-Lowry theory states that acid-base reactions are a competition for a proton. For example, take a look at the reaction of ammonia with water:

$$NH_3(g) + H_2O(l) \rightleftarrows NH_4^+ + OH^-$$

Ammonia is a base (it accepts the proton), and water is an acid (it donates the proton) in the forward (left to right) reaction. But in the reverse reaction (right to left), the ammonium ion is an acid and the hydroxide ion is a base.

If water is a stronger acid than the ammonium ion, then there is a relatively large concentration of ammonium and hydroxide ions at equilibrium. If, however, the ammonium ion is a stronger acid, much more ammonia than ammonium ion is present at equilibrium.

Brønsted and Lowry said that an acid reacts with a base to form *conjugate acid-base pairs*. Conjugate acid-base pairs differ by a single H⁺. NH_3 is a base, for example, and NH_4^+ is its conjugate acid. H_2O is an acid in the reaction between ammonia and water, and OH⁻ is its conjugate base. In this reaction, the hydroxide ion is a strong base and ammonia is a weak base, so the equilibrium is shifted to the left — there's not much hydroxide at equilibrium.

Acting as either an acid or base: Amphoteric water

Water can act as either an acid or a base, depending on what it's combined with. Substances that can act as either an acid or a base are called *amphoteric*. If you put water with an acid, it acts as a base, and vice versa. For instance, when acetic acid reacts with water, water acts as a base, or a proton acceptor. But in the reaction with ammonia, water acts as an acid, or a proton donor. (See the earlier section "Weak: Ionizing partially" for details on both reactions.)

But can water react with itself? Yes, it can. Two water molecules can react with each other, with one donating a proton and the other accepting it:

$$H_2O(l) + H_2O(l) \rightleftharpoons H_3O^+ + OH^-$$

This reaction is an equilibrium reaction. A modified equilibrium constant, called the K_w (which stands for *water dissociation constant*), is associated with this reaction. The K_w has a value of 1.0×10^{-14} and has the following form:

$$1.0 \times 10^{-14} = K_w = [H_3O+] [OH^-]$$

In pure water, the $[H_3O^+]$ equals the $[OH^-]$ from the balanced equation, so $[H_3O^+] = [OH^-] = 1.0 \times 10^{-7}$. The K_w value is a constant.

This value allows you to convert from $[H^+]$ to $[OH^-]$, and vice versa, in *any* aqueous solution, not just pure water. In aqueous solutions, the hydronium ion and hydroxide ion concentrations are rarely going to be equal. But if you know one of them, K_w allows you to figure out the other one.

Take a look at the 2.0 M acetic acid solution problem in the section "Acetic acid and other weak acids," earlier in this chapter. You find that the $[H_3O^+]$ is 6.0×10^{-3}. Now you have a way to calculate the $[OH^-]$ in the solution by using the K_w relationship:

$$K_w = 1.0 \times 10^{-14} = \left[H_3O^+ \right]\left[OH^- \right]$$

$$1.0 \times 10^{-14} = \left[6.0 \times 10^{-3} \right]\left[OH^- \right]$$

$$\frac{1.0 \times 10^{-14}}{6.0 \times 10^{-3}} = \left[OH^- \right]$$

$$1.7 \times 10^{-12} = \left[OH^- \right]$$

Showing True Colors with Acid-Base Indicators

Indicators are substances (organic dyes) that change color in the presence of an acid or base. You may be familiar with an acid-base indicator plant — the hydrangea. If it's grown in acidic soil, it turns pink; if it's grown in alkaline soil, it turns blue.

In chemistry, indicators are used to indicate the presence of an acid or base. Chemists have many indicators that change at slightly different pHs, but the two indicators used most often are litmus paper and phenolphthalein. I discuss both in this section.

Doing a quick color test with litmus paper

Litmus is a substance that is extracted from a type of lichen and absorbed into porous paper. There are three different types of litmus — red, blue, and neutral. Red litmus is used to test for bases, blue litmus is used to test for acids, and neutral litmus can be used to test for both. Here's how the paper reacts to acids and bases:

✔ If a solution is acidic, both blue and neutral litmus will turn red.

✔ If a solution is basic, both red and neutral litmus will turn blue.

Litmus paper is a good, quick test for acids and bases.

Phenolphthalein: Finding concentration with titration

Phenolphthalein (pronounced *fe-nul-tha-Leen*) is a commonly used indicator. Phenolphthalein is

✔ Clear and colorless in an acid solution

✔ Pink in a basic solution

Chemists use phenolphthalein in a procedure called a *titration*, in which they determine the concentration of an acid or base by its reaction with a base or acid of known concentration (see Chapter 10 for info on molarity and other solution concentration units). Here's how to evaluate an acid solution using titration:

1. **Add a couple drops of phenolphthalein to a known volume of the acid solution you want to test.**

   Because you're adding the indicator to an acidic solution, the solution in the flask remains clear and colorless.

   Suppose, for example, that you want to determine the molar concentration of an HCl solution. First, you place a known volume (say, 25.00 milliliters measured accurately with a pipette) in an Erlenmeyer flask (that's just a flat-bottomed, conical-shaped flask) and add a couple drops of phenolphthalein solution.

2. **Add small, measured amounts of a base of known molarity (concentration) until the solution turns light pink.**

   Add small amounts of a standardized sodium hydroxide (NaOH) solution of known molarity (for example, 0.100 M) with a buret. (A *buret* is a graduated glass tube with a small opening and a stopcock valve, which helps you measure precise volumes of solution.) Keep adding base until the solution turns the faintest shade of pink detectable. I call this the *endpoint* of the titration, the point at which the acid has been exactly neutralized by the base.

3. **Write the balanced equation for the reaction.**

   Here's the reaction:

   $$HCl(aq) + NaOH(aq) \rightarrow H_2O(l) + NaCl(aq)$$

4. **Calculate the molarity of the acidic solution.**

   From the balanced equation, you can see that the acid and base react in a 1:1 mole ratio. So if you can calculate the moles of bases added, you'll also know the number of moles of HCl present. Suppose that it takes 35.50 milliliters of the 0.100 M NaOH to reach the endpoint of the titration of the 25.00 milliliters of the HCl solution. Knowing the volume of the acid solution then allows you to calculate the molarity (note that you convert the milliliters to liters so that your units cancel nicely):

   $$\frac{0.100 \text{ mol NaOH}}{L} \cdot \frac{0.03550 \text{ L}}{1} \cdot \frac{1 \text{ mol HCl}}{1 \text{ mol NaOH}}$$

   $$\cdot \frac{1}{0.02500 \text{ L}} = 0.142 \text{ M HCl}$$

You can calculate the titration of a base with a standard acid solution (one of known concentration) in exactly the same way, except the endpoint is the first disappearance of the pink color.

Phun with the pH Scale

The acidity of a solution is related to the concentration of the hydronium ion in the solution: The more acidic the solution is, the higher the concentration. In other words, a solution in which the $[H_3O^+]$ equals 1.0×10^{-2} is more acidic than a solution in which the $[H_3O^+]$ equals 1.0×10^{-7}.

Scientists developed the *pH scale*, a scale based on the $[H_3O^+]$, to more easily tell, at a glance, the relative acidity of a solution. The *pH* is defined as the negative logarithm (abbreviated as *log*) of $[H_3O^+]$. Mathematically, it looks like this:

$$pH = - \log [H_3O^+]$$

In pure water, the $[H_3O^+]$ equals 1.0×10^{-7}, based on the water dissociation constant, K_w (see "Acting as either an acid or a base: Amphoteric water," earlier in this chapter. Using this mathematical relationship, you can calculate the pH of pure water:

$$pH = -\log [H_3O^+]$$
$$pH = -\log [1.0 \times 10^{-7}]$$
$$pH = -[-7]$$
$$pH = 7$$

The pH of pure water is 7. Chemists call this point on the pH scale *neutral*. A solution is *acidic* if it has a larger $[H_3O^+]$ than water and a smaller pH value than 7. A *basic* solution has a smaller $[H_3O^+]$ than water and a larger pH value than 7.

The pH scale really has no end. You can have a solution of pH that registers less than 0. (A 10 M HCl solution, for example, has a pH of –1.) However, the 0 to 14 range is a convenient range to use for weak acids and bases and for dilute solutions of strong acids and bases. Figure 11-2 shows the pH scale.

The $[H_3O^+]$ of a 2.0 M acetic acid solution is 6.0×10^{-3}. Looking at the pH scale, you see that this solution is acidic. Now calculate the pH of this solution:

$$pH = -\log [H_3O^+]$$
$$pH = -\log [6.0 \times 10^{-3}]$$
$$pH = -[-2.22]$$
$$pH = 2.22$$

The *pOH* is the negative logarithm of the $[OH^-]$, and it can be useful in calculating the pH of a solution. You can calculate the pOH of a solution just like the pH by taking the negative log of the hydroxide ion concentration. If you use the K_w expression (which enables you to calculate the $[H_3O^+]$ if you have the $[OH^-]$; see the section "Acting as either an acid or base: Amphoteric water") and take the negative log of both sides, you get 14 = pH + pOH. This equation makes it easy to go from pOH to pH.

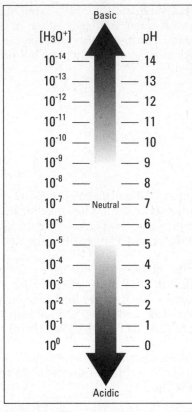

Figure 11-2: The pH scale.

Just as you can you convert from $[H_3O^+]$ to pH, you can also go from pH to $[H_3O^+]$. To do this, you use what's called the *antilog relationship*, which is

$$[H_3O^+] = 10^{-pH}$$

Human blood, for example, has a pH of 7.3. Here's how you calculate the $[H_3O^+]$ from the pH of blood:

$$[H_3O^+] = 10^{-pH}$$
$$[H_3O^+] = 10^{-7.3}$$
$$[H_3O^+] = 5.01 \times 10^{-8}$$

You can use the same procedure to calculate the $[OH^-]$ from the pOH.

Chapter 12

Clearing the Air on Gases

● ●

In This Chapter

▶ Accepting the Kinetic Molecular Theory of Gases

▶ Understanding the gas laws

● ●

Gases are all around you. Because gases are generally invisible, you may not think of them directly, but you're certainly aware of their properties. You breathe a mixture of gases that you call air. You check the pressure of your automobile tires, and you check the atmospheric pressure to see whether a storm is coming. You burn gases in your gas grill and lighters. You fill birthday balloons for your loved ones.

In this chapter, I introduce you to gases at both the microscopic and macroscopic levels. I show you one of science's most successful theories: the Kinetic Molecular Theory of Gases. I explain the macroscopic properties of gases and show you the important interrelationships among them. I also show you how these relationships come into play in the calculations of chemical reactions involving gases. This chapter is a real gas!

The Kinetic Molecular Theory: Assuming Things about Gases

A theory is useful to scientists if it describes the physical system they're examining and allows them to predict what will happen if they change some variable. The Kinetic Molecular Theory of Gases does just that. It has limitations (all theories do), but it's one of the most useful theories in chemistry. Here are the theory's basic *postulates* — assumptions, hypotheses, axioms (pick your favorite word) you can accept as being consistent with what you observe in nature:

- **Postulate 1: Gases are composed of tiny particles, either atoms or molecules.** Unless you're discussing matter at really high temperatures, the particles referred to as gases tend to be relatively small. The more-massive particles clump together to form liquids or even solids, so gas particles are normally small with relatively low atomic and molecular weights.

- **Postulate 2: The gas particles are so small when compared to the distances between them that the volume the gas particles themselves take up is negligible and is assumed to be zero.** Individual gas particles do take up some volume — that's one of the properties of matter. But if the gas particles are small (which they are), and there aren't many of them in a container, you say that their volume is *negligible* when compared to the volume of the container or the space between the gas particles. Sure, they have a volume, but it's so small that it's insignificant (just like a dollar on the street doesn't represent much at all to a multimillionaire; it may as well be a piece of scrap paper).

  This explains why gases are compressible. There's a lot of space between the gas particles, so you can squeeze them together. This isn't true in solids and liquids, where the particles are *much* closer together.

- **Postulate 3: The gas particles are in constant random motion, moving in straight lines and colliding with the inside walls of the container.** The gas particles are always moving in a straight-line motion. They continue to move in these straight lines until they collide with something, either each other or the inside walls of the container. The particles also all move in different directions, so the collisions with the inside walls of the container tend to be uniform over the entire inside surface. A balloon, for instance, is relatively spherical because the gas particles are hitting all points of the inside walls the same. The collision of the gas particles with the inside walls of the container is called *pressure*.

This postulate explains why gases uniformly mix if you put them in the same container. It also explains why, when you drop a bottle of cheap perfume at one end of the room, the people at the other end of the room are able to smell it right away.

✓ **Postulate 4: The gas particles are assumed to have negligible attractive or repulsive forces between each other.** In other words, you assume that the gas particles are totally independent, neither attracting nor repelling each other. That said, if this assumption were correct, chemists would never be able to liquefy a gas, which they can. However, the attractive and repulsive forces are generally so small that you can safely ignore them.

The assumption is most valid for nonpolar gases, such as hydrogen and nitrogen, because the attractive forces involved are *London forces,* weak forces that have to do with the ebb and flow of the electron orbitals. However, if the gas molecules are polar, as in water and HCl, this assumption can become a problem, because the forces are stronger. (Turn to Chapter 6 for the scoop on London forces and polar things — all related to the attraction between molecules.)

✓ **Postulate 5: The gas particles may collide with each other. These collisions are assumed to be *elastic,* with the total amount of kinetic energy of the two gas particles remaining the same.** When gas particles hit each other, no *kinetic energy* — energy of motion — is lost. That is, the type of energy doesn't change; the particles still use all that energy for movement. However, kinetic energy may be transferred from one gas particle to the other. For example, imagine two gas particles — one moving fast and the other moving slow — colliding. The one that's moving slow bounces off the faster particle and moves away at a greater speed than before, and the one that's moving fast bounces off the slower particle and moves away at a slower speed. But the sum of their kinetic energy remains the same.

✓ **Postulate 6: The Kelvin temperature is directly proportional to the *average* kinetic energy of the gas particles.** The gas particles aren't all moving with the same amount of kinetic energy. A few are moving relatively slow, and a few are moving very fast, but most are somewhere in between these two extremes. Temperature, particularly as measured using the Kelvin temperature scale, is directly related to the *average* kinetic energy of the gas. If you heat the gas so that the Kelvin temperature (K) increases, the average kinetic energy of the gas also increases. (*Note:* To calculate the Kelvin temperature, add 273 to the Celsius temperature: K = °C + 273. Temperature scales and average kinetic energy are all tucked neatly into Chapter 1.)

A gas that obeys all the postulates of the Kinetic Molecular Theory is called an *ideal gas.* Obviously, no real gas obeys the assumptions made in the second and fourth postulates *exactly.* But a nonpolar gas at high temperatures and low pressure (concentration) approaches ideal gas behavior.

Relating Physical Properties with Gas Laws

Various scientific laws describe the relationships between four important physical properties of gases:

- ✔ Volume

- ✔ Pressure

- ✔ Temperature (in Kelvin units)

- ✔ Amount

This section covers those various laws. Boyle's, Charles's, and Gay-Lussac's laws each describe the relationship between two properties while keeping the other two properties constant; in other words, you take two properties, change one, and then see its effect on the second. Another law — a combo of Boyle's, Charles's, and Gay-Lussac's individual laws — enables you to vary more than one property at a time.

That combo law doesn't let you vary the physical property of amount. Avogadro's Law, however, does. And there's even an ideal gas law, which lets you take into account variations in all four physical properties.

Boyle's law: Pressure and volume

Boyle's law (named after Robert Boyle, a 17th-century English scientist) describes the pressure-volume relationship of gases if you keep the temperature and amount of the gas constant. The law states that there's an inverse relationship between the volume and air *pressure* (the collision of the gas particles with the inside walls of the container): As the volume decreases, the pressure increases, and vice versa. He determined that the product of the pressure and the volume is a constant *(k):*

$$PV = k$$

Suppose you have a cylinder that contains a certain volume of gas at a certain pressure. When you decrease the volume, the same number of gas particles is now contained in a much smaller space, and the number of collisions increases significantly. Therefore, the pressure is greater.

Now consider a case where you have a gas at a certain pressure (P_1) and volume (V_1). If you change the volume to some new value (V_2), the pressure also changes to a new value (P_2). You can use Boyle's Law to describe both sets of conditions:

$$P_1V_1 = k$$
$$P_2V_2 = k$$

The constant, k, is going to be the same in both cases, so you can say the following, if the temperature and amount of gas don't change:

$$P_1V_1 = P_2V_2$$

This equation is another statement of Boyle's Law — and it's really a more useful one, because you normally deal with changes in pressure and volume.

If you know three of the preceding quantities, you can calculate the fourth one. For example, suppose that you have 5.00 liters of a gas at 1.00 atm pressure, and then you decrease the volume to 2.00 liters. What's the new pressure? Use the formula. Substitute 1.00 atm for P_1, 5.00 liters for V_1, and 2.00 liters for V_2, and then solve for P_2:

$P_1V_1 = P_2V_2$

$(1.00 \text{ atm})(5.00 \text{ L}) = P_2(2.00 \text{ L})$

$P_2 = (1.00 \text{ atm})(5.00 \text{ L})/2.00 \text{ L} = 2.50 \text{ atm}$

The answer makes sense; you decreased the volume, and the pressure increased, which is exactly what Boyle's law says.

Charles's law: Volume and temperature

Charles's law (named after Jacques Charles, a 19th-century French chemist) has to do with the relationship between volume and temperature, keeping the pressure and amount of the gas constant. Ever leave a bunch of balloons in a hot car while running an errand? Did you notice that they expanded when you returned to the car?

Charles's law says that the volume is directly proportional to the Kelvin temperature. Mathematically, the law looks like this:

$V = bT$ or $V/T = b$ (where b is a constant)

This is a direct relationship: As the temperature increases, the volume increases, and vice versa. For example, if you placed a balloon in the freezer, the balloon would get smaller. Inside the freezer, the external pressure, or atmospheric pressure, is the same, but the gas particles inside the balloon aren't moving as fast, so the volume shrinks to keep the pressure constant. If you heat the balloon, the balloon expands and the volume increases.

If the temperature of a gas with a certain volume (V_1) and Kelvin temperature (T_1) is changed to a new Kelvin temperature (T_2), the volume also changes (V_2):

$V_1/T_1 = b$ $V_2/T_2 = b$

The constant, b, is the same, so

$V_1/T_1 = V_2/T_2$ (with the pressure and amount of gas held constant and temperature expressed in K)

If you have three of the quantities, you can calculate the fourth. For example, suppose you live in Alaska and are outside in the middle of winter, where the temperature is –23°C. You blow up a balloon so that it has a volume of 1.00 liter. You then take it inside your home, where the temperature is a toasty 27°C. What's the new volume of the balloon?

First, convert your temperatures to Kelvin by adding 273 to the Celsius temperature:

Inside: –23°C + 273 = 250 K

Outside: 27°C + 273 = 300 K

Now you can solve for V_2, using the following setup:

$$V_1/T_1 = V_2/T_2$$

Multiply both sides by T_2 so that V_2 is on one side of the equation by itself:

$$[V_1 T_2]/T_1 = V_2$$

Then substitute the values to calculate the following answer:

$$[(1.00\ \text{L})(300\ \text{K})]/250\ \text{K} = V_2 = 1.20\ \text{L}$$

It's a reasonable answer, because Charles's Law says that if you increase the Kelvin temperature, the volume increases.

Gay-Lussac's Law: Pressure and temperature

Gay-Lussac's Law (named after the 19th-century French scientist Joseph-Louis Gay-Lussac) deals with the relationship between the pressure and temperature of a gas if its volume and amount are held constant. Imagine, for example, that you have a metal tank of gas. The tank has a certain volume, and the gas inside has a certain pressure. If you heat the tank, you increase the kinetic energy of the gas particles. So they're now moving much faster, and they're hitting the inside walls of the tank not only more often but also with more force. The pressure has increased.

Gay-Lussac's Law says that the pressure is directly proportional to the Kelvin temperature. Mathematically, Gay-Lussac's Law looks like this:

$$P = kT \text{ (or } P/T = k \text{ at constant volume and amount)}$$

Consider a gas at a certain Kelvin temperature and pressure (T_1 and P_1), with the conditions being changed to a new temperature and pressure (T_2 and P_2):

$$P_1/T_1 = P_2/T_2$$

If you have a tank of gas at 800 torr pressure and a temperature of 250 Kelvin, and it's heated to 400 Kelvin, what's the new pressure? Starting with $P_1/T_1 = P_2/T_2$, multiply both sides by T_2 so you can solve for P_2:

$$[P_1 T_2]/T_1 = P_2$$

Now substitute the values to calculate the following answer:

$$P_2 = [(800 \text{ torr})(400 \text{ K})]/250 \text{ K} = 1{,}280 \text{ torr}$$

This is a reasonable answer because if you heat the tank, the pressure should increase.

The combined gas law: Pressure, volume, and temp.

You can combine Boyle's Law, Charles's Law, and Gay-Lussac's Law into one equation to handle situations in which two or even three gas properties change. Trust me, you don't want me to show you exactly how it's done, but the end result is called the *combined gas law,* and it looks like this:

$$P_1 V_1/T_1 = P_2 V_2/T_2$$

P is the pressure of the gas (in atm, mm Hg, torr, and so on), V is the volume of the gas (in appropriate units), and T is the temperature (in Kelvin). The $_1$ and $_2$ stand for the initial and final conditions, respectively. The amount of gas is still held constant: No gas is added, and no gas escapes. There are six quantities involved in this combined gas law; knowing five allows you to calculate the sixth.

For example, suppose that a weather balloon with a volume of 25.0 liters at 1.00 atm pressure and a temperature of 27°C is allowed to rise to an altitude where the pressure is 0.500 atm and the temperature is –33°C. What's the new volume of the balloon?

Before working this problem, do a little reasoning. The temperature is decreasing, so that should cause the volume to decrease (Charles's Law). However, the pressure is also decreasing, which should cause the balloon to expand (Boyle's Law). These two factors are competing, so at this point, you don't know which will win out.

You're looking for the new volume (V_2), so rearrange the combined gas law to obtain the following equation (by multiplying each side by T_2 and dividing each side by P_2, which puts V_2 by itself on one side):

$$[P_1 V_1 T_2]/[P_2 T_1] - V_2$$

Now identify your quantities:

$P_1 = 1.00$ atm; $V_1 = 25.0$ liters; $T_1 = 27°C + 273 = 300.$ K
$P_2 = 0.500$ atm; $T_2 = -33°C + 273 = 240.$ K

Now substitute the values to calculate the following answer:

$V_2 = [(1.00$ atm$)(25.0$ L$)(240.$ K$)]/[(0.500$ atm$)(300.$ K$)] = 10.0$ L

Because the volume increased overall in this case, Boyle's Law had a greater effect than Charles's Law.

Avogadro's Law: The amount of gas

Amedeo Avogadro (the same Avogadro who gave us his famous number of particles per mole — see Chapter 9) determined, from his study of gases, that equal volumes of gases at the same temperature and pressure contain equal numbers of gas particles. So *Avogadro's law* says that the volume of a gas is directly proportional to the number of moles of gas (number of gas particles) at a constant temperature and pressure. Mathematically, Avogadro's law looks like this:

$$V = kn \quad \text{(at constant temperature and pressure)}$$

In this equation, k is a constant and n is the number of moles of gas. If you have a number of moles of gas (n_1) at one volume (V_1), and the moles change due to a reaction (n_2), the volume also changes (V_2), giving you the equation

$$V_1/n_1 = V_2/n_2$$

A very useful consequence of Avogadro's Law is that you can calculate the volume of a mole of gas at any temperature and pressure. An extremely useful form to know when calculating the volume of a mole of gas is that 1 mole of any gas at STP occupies 22.4 liters. *STP* in this case is not an oil or gas additive; it stands for *standard temperature and pressure.*

✔ **Standard pressure:** 1.00 atm (760 torr or mm Hg)

✔ **Standard temperature:** 273 K

This relationship between moles of gas and liters gives you a way to convert the gas from a mass to a volume. For example, suppose that you have 50.0 grams of oxygen gas (O_2) and you want to know its volume at STP. You can set up the problem like this (see Chapters 9 and 10 for the nuts and bolts of using moles in chemical equations):

$$\frac{50.0 \text{g } O_2}{1} \times \frac{1 \text{ mol } O_2}{32.0 \text{g}} \times \frac{22.4 \text{L}}{1 \text{ mol } O_2} = 35.0 \text{ L}$$

You now know that the 50.0 grams of oxygen gas occupies a volume of 35.0 liters at STP.

If the gas isn't at STP, you can use the combined gas law (from the preceding section) to find the volume at the new pressure and temperature — or you can use the ideal gas equation, which I show you next.

The ideal gas equation: Putting it all together

If you take Boyle's law, Charles's law, Gay-Lussac's law, and Avogadro's law and throw them into a blender, turn the blender on high for a minute, and then pull them out, you get the *ideal gas equation* — a way of working in volume, temperature, pressure, and amount of a gas. The ideal gas equation has the following form:

$$PV = nRT$$

The P represents pressure in atmospheres (atm), the V represents volume in liters (L), the n represents moles of gas, the T represents the temperature in Kelvin (K), and the R represents the *ideal gas constant*, which is 0.0821 liters atm/K-mol.

Using the value of the ideal gas constant, the pressure must be expressed in atm, and the volume must be expressed in liters. You can calculate other ideal gas constants if you really want to use torr and milliliters, for example, but why bother? It's easier to memorize one value for R and then remember to express the pressure and volume in the appropriate units. Naturally, you'll always express the temperature in Kelvin when working any kind of gas law problem.

The ideal gas equation gives you an easy way to convert a gas from a mass to a volume if the gas is not at STP. For instance, what's the volume of 50.0 grams of oxygen at 2.00 atm and 27.0°C? The first thing you have to do is convert the 50.0 grams of oxygen to moles using the molecular weight of O_2:

(50.0 grams) • (1 mol/32.0 grams) = 1.562 mol

Now take the ideal gas equation and rearrange it so you can solve for V:

$$PV = nRT$$

$$V = nRT/P$$

Add your known quantities to calculate the following answer:

V = [(1.562 mol) • (0.0821 L atm/K-mol) • (300 K)]/ 2.00 atm = 19.2 L

Chapter 13

Ten Serendipitous Discoveries in Chemistry

In This Chapter

▶ Reviewing some great discoveries

▶ Examining some famous people of science

Chemistry doesn't always go as planned. This chapter presents ten stories of good scientists who discovered something they didn't know they were looking for.

Archimedes: Streaking Around

Archimedes was a Greek mathematician who lived in the third century BCE. Hero, the king of Syracuse, gave Archimedes the task of determining whether Hero's new gold crown was composed of pure gold, which it was supposed to be, or whether the jeweler had substituted an alloy and pocketed the extra gold. Archimedes figured that if he could measure the density of the crown and compare it to that of pure gold, he'd know whether the jeweler had been dishonest. But although he knew how to measure the weight of the crown, he couldn't figure out how to measure its volume in order to get the density.

Needing some relaxation, he decided to bathe at the public baths. As he stepped into the full tub and saw the water overflow, he realized that the volume of his body that was submerged was equal to the volume of water that overflowed. He had his answer for measuring the volume of the crown. Legend has it that he got so excited that he ran home naked through the streets, yelling, "Eureka, eureka!" (I've found it!).

Vulcanization of Rubber

Rubber, in the form of latex, was discovered in the early 16th century in South America, but it gained little acceptance because it became sticky and lost its shape in the heat. Charles Goodyear was trying to find a way to make the rubber stable when he accidentally spilled a batch of rubber mixed with sulfur on a hot stove. He noticed that the resulting compound didn't lose its shape in the heat. Goodyear went on to patent the *vulcanization process,* the chemical process used to treat crude or synthetic rubber or plastics to give them useful properties such as elasticity, strength, and stability.

Molecular Geometry

In 1884, the French wine industry hired Louis Pasteur to study a compound left on wine casks during fermentation — racemic acid. Pasteur knew that racemic acid was identical to tartaric acid, which was known to be *optically active* — that is, it rotated polarized light in one direction or another. When Pasteur examined the salt of racemic acid under a microscope, he noticed that two types of crystals were present and that they were mirror images of each other. Using a pair of tweezers, Pasteur laboriously separated the two types of crystals and determined that they were both optically active, rotating polarized light the same amount but in different directions. This discovery opened up a new area of chemistry and showed how important molecular geometry is to the properties of molecules.

Mauve Dye

In 1856, William Perkin, a student at The Royal College of Chemistry in London, decided to stay home during the Easter break and work in his lab on the synthesis of quinine. (I guarantee you that working in the lab isn't what my students do during their Easter break!) During the course of his experiments, Perkin created some black gunk. As he was cleaning the reaction flask with alcohol, he noticed that the gunk dissolved and turned the alcohol purple — mauve, actually. This was the synthesis of the first artificial dye.

Kekulé: The Beautiful Dreamer

Friedrich Kekulé, a German chemist, was working on the structural formula of benzene, C_6H_6, in the mid-1860s. Late one night, he was sitting in his apartment in front of a fire. He began dozing off and, in the process, saw groups of atoms dancing in the flames like snakes. Then, suddenly, one of the snakes reached around and made a circle, or a ring. This vision startled Kekulé to full consciousness, and he realized that benzene had a ring structure. Kekulé's model for benzene paved the way for the modern study of aromatic compounds.

Discovering Radioactivity

In 1856, Henri Becquerel was studying the *phosphorescence* (glowing) of certain minerals when exposed to light. In his experiments, he'd take a mineral sample, place it on top of a heavily wrapped photographic plate, and expose it to strong sunlight. He was preparing to conduct one of these experiments when a cloudy spell hit Paris. Becquerel put a mineral sample on top of the plate and put it in a drawer for safekeeping. Days later, he went ahead and developed the photographic plate and, to his surprise, found the brilliant image of the crystal, even though it hadn't been exposed to light. The mineral sample contained uranium. Becquerel had discovered radioactivity.

Finding Really Slick Stuff: Teflon

Roy Plunkett, a DuPont chemist, discovered Teflon in 1938. He was working on the synthesis of new refrigerants. He had a full tank of tetrafluoroethylene gas delivered to his lab, but when he opened the valve, nothing came out. He wondered what had happened, so he cut the tank open. He found a white substance that was very slick and nonreactive. The gas had polymerized into the substance now called Teflon. It was used during World War II to make gaskets and valves for the atomic bomb processing plant. After the war, Teflon finally made its way into the kitchen as a nonstick coating for frying pans.

Stick 'Em Up! Sticky Notes

In the mid-1970s, a chemist by the name of Art Frey was working for 3M in its adhesives division. Frey, who sang in a choir, used little scraps of paper to keep his place in his choir book, but they kept falling out. At one point, he remembered an adhesive that had been developed but rejected a couple years earlier because it didn't hold things together well. The next Monday, he smeared some of this "lousy" adhesive on a piece of paper and found that it worked very well as a bookmark — and it peeled right off without leaving a residue. Thus was born those little yellow sticky notes you now find posted everywhere.

Growing Hair

In the late 1970s, minoxidil, patented by Upjohn, was used to control high blood pressure. In 1980, Dr. Anthony Zappacosta mentioned in a letter published in *The New England Journal of Medicine* that one of his patients using minoxidil for high blood pressure was starting to grow hair on his nearly bald head. Dermatologists took note, and one — Dr. Virginia Fiedler-Weiss — crushed up some of the tablets and made a solution that some of her patients applied topically. It worked in enough cases that you now see Minoxidil as an over-the-counter hair-growth medicine.

Sweeter Than Sugar

In 1879, a chemist by the name of Fahlberg was working on a synthesis problem in the lab. He accidentally spilled on his hand one of the new compounds he'd made, and he noticed that it tasted sweet. He called this new substance *saccharin*. James Schlatter discovered the sweetness of *aspartame* while working on a compound used in ulcer research. He accidentally got a bit of one of the esters he'd made on his fingers. He noticed its sweetness when he licked his fingers while picking up a piece of paper.

Index